Assembling a Solar Generator:

How to Harness the Sun for Power when you Need it Most

by Jay Warmke

Published by BRS Press
ISBN-13: 978-0-9791611-4-8
ISBN-10: 0-9791611-4-2
Text © 2016 Jay & Annie Warmke
Graphs & Charts (unless otherwise noted) © 2016 Jay & Annie Warmke
Layout and Design by BRS Press
A special thanks to:
Annie Warmke, Kevin Billman - *Editors*
Clintonville Energy Co-Op - *Guinea Pigs/Editors*

Contact us at:

Blue Rock Station
1190 Virginia Ridge Road
Philo, Ohio 43771 USA
Telephone: +1 (740) 674- 4300, Email: annie@bluerockstation.com
www.bluerockstation.com

BRS Press

Dedication

*To my friend Tim Chavez,
who constantly reminds me
just how cool solar technology is,
and who helped build our very first
solar generator out of parts he
salvaged from parking meters
(don't ask).*

About the Author

Jay Warmke is co-owner of Blue Rock Station (www.bluerockstation.com), a sustainable living center that features the first Earthship (a 2,200 sq ft passive solar home constructed out of garbage) built east of the Mississippi, many straw bale buildings, a plastic bottle greenhouse, a solar shower, gardens, milk goats, lots of summer interns, and way too many cats. In the years since establishing Blue Rock Station, more than 30,000 people have wandered through their living room – drawn by the vision of a truly sustainable lifestyle.

Jay is the author of numerous green technology books and articles, including *Green Technology: Principles and Practices* (Marcraft), *Solar Photovoltaic Installer Certification Guide* (Marcraft), *Wind Turbine Installer Certification Guide* (Marcraft), *When the Biomass Hits the Wind Turbine* (BRS Press), *Understanding Photovoltaics: A Study Guide for Solar Electric Certification Programs* (BRS Press), and many more.

He is vice president of the board of directors of Green Energy Ohio, and vice president of the International Certification and Accreditation Council. He has served as a committee member on ETA's renewable energy committee since 2010 and as chairman sin 2015. In 2015 he was elected to ETA's Board of Directors. He is on the advisory council of Ohio State University's renewable energy program, as well as an adviser to Tolles Career & Technical Center renewable energy program. He is a former instructor of Renewable Energy at Central Ohio Technical College, current instructor at Zane State College and Jay also served as National Chairman: SkillsUSA Sustainable Solutions Contest (2010-2014).

Along with Annie Warmke, he was presented the 2009 Sustainable Living Award from Rural Action Ohio, and Green Energy Ohio named him "Pioneer of the Year" in 2011. He was also named the Electronic Technician's Association Educator of the Year for 2013.

Table of Contents

- Chapter 1: Introduction 1
- Chapter 2: In Case of Emergency 5
- Chapter 3: Energy Efficiency 9
- Chapter 4: Generator Parts 11
- Chapter 5: Selecting the Inverter 17
- Chapter 6: The Battery Bank 29
- Chapter 7: The Solar Panels 43
- Chapter 8: The Charge Controller 55
- Chapter 9: Wiring the Generator 61
- Chapter 10: Safety 77
- Chapter 11: Assembling a Solar Generator 81
- Chapter 12: A Picture is Worth 1,000 Words 95
- Chapter 13: Maintenance 109
- Chapter 14: Troubleshooting 111
- Figures & Tables 115
- Index: 117

Chapter One
Introduction

Each and every year the US experiences over 3,000 power outages. Tens of millions of people are effected, and it has been estimated that these power disruptions cost the economy over $150 billion annually due to property damage, as well as lost revenue and wages (*Eaton Power Outage Annual Report 2013*).

According to a recent White House statement (2013), investment in the nation's electrical grid is inadequate. More than 75% of the current grid infrastructure is over 25 years old. The report noted that severe weather is likely to become more frequent due to climate change, and concluded by emphasizing that, "Developing a smarter, more resilient electric grid is one step that can be taken now to ensure the welfare of the millions of current and future Americans who depend on the grid for reliable power."

A 2013 Associated Press report emphasized the sorry state of the grid, noting that not only are power outages becoming more frequent, but that they take on average 20% longer to resolve than just a decade earlier. While many outages take days to resolve, the average outage

is 197 minutes (just over 3 hours). It is estimated that every single day over 500,000 Americans experience a power outage of about two hours.

Assuming that corporate and political powers can even muster the economic and political will to address this growing problem – it will take decades and billions of dollars to fully resolve. In the meantime, most industry experts expect that the situation will get much worse before it gets better.

Another contributing factor to the potential destruction associated with power outages is that they are often associated with severe weather events. Take for example, the top three most severe outages of 2013.

Three Most Significant Outages of 2013:

1. On December 21st, an ice storm hit Michigan, leaving more than ½ million residents without power. More than 150,000 were still without power on Christmas Day, and all power was not fully restored until December 28th. Not only were the holidays interrupted for many thousands – imagine the burst pipes and damaged homes that resulted from the outage.
2. More freezing damage was the result of a deadly blizzard that pounded the Northeast US on Feb. 8th, bringing more than three feet of snow to some areas and cutting power to 650,000 homes and businesses, including 350,000 across Rhode Island.
3. Massive rainfall and flooding caused outages in Ontario, Canada on July 8th, leaving 560,000 customers without power. Those that relied on pumps to keep their homes and basements dry were simply out of luck.

CHAPTER 1: INTRODUCTION

Top 10 States Affected by Outages:*

2013	2012	2011
1. California (464)	1. California (510)	1. California (371)
2. Texas (159)	2. New York (133)	2. New York (159)
3. Michigan (153)	3. Texas (131)	3. Texas (153)
4. Pennsylvania (144)	4. Michigan (125)	4. Michigan (143)
5. Ohio (136)	5. New Jersey (119)	5. Penn. (134)
6. New York (125)	6. Pennsylvania (109)	6. Illinois (129)
7. Virginia (117)	7. Ohio (91)	7. Ohio (121)
8. New Jersey (116)	8. Washington (90)	8. New Jersey (107)
9. Washington (104)	9. Illinois (76) tie	9. Washington (91)
10. Mass. (98)	9. Virginia (76) tie	10. Wisconsin (89)

*from Eaton Power Outage Annual Report 2013

Thunderstorms and severe cold weather are often responsible for power outages. Outages that occur just when you need the electricity most to protect your home and family.

ASSEMBLING A SOLAR GENERATOR

Some Basic Electrical Terms:

Before we learn how solar generators operate, and how to design and build one, you will need to understand a few electrical terms. This does not mean you need to be an electrician - but a little bit of electrical knowledge is a must if you are to proceed with this project.

Volts (V): This is one of those terms that people use all the time and kind of sort of understand it, but it is very difficult to define without your eyes glazing over. Essentially voltage is the difference in electrical potential between two points, typically connected by a conductive wire. Not much help.

It might be easier to think of voltage using the classic water analogy. Using this analogy, a volt can be thought of as the water pressure (how fast the water will flow when there is a current present). The pressure exists whether water is flowing or not.

Amps (I): An amp is the current, or flow of electricity. Using the water analogy, amps can be thought of as the size of the pipe. A circuit with high amps and low voltage is like a large pipe with water moving rather slowly. Low amps with high voltage is like a smaller pipe with water flowing very quickly.

Watts (W): This is the amount of power produced. Simply stated, watts = amps x volts (or $W = I * V$).

Watt-hour (Wh): When we buy electricity, it is measured in watt-hours (Wh) or more specifically in **kilowatt-hours (kWh)**. A watt-hour is one watt of power consumed at a constant rate for a period of one hour. A kilowatt-hour is 1,000 watts of power consumed at a constant rate for a period of one hour.

Chapter Two
In case of emergency

Assuming we accept that climate change and aging infrastructure will result in more severe and more frequent power outages. What can we do about it?

Well, as any good Boy or Girl Scout knows – be prepared. But what does this entail?

The American Red Cross has some suggestions on how to prepare for a power outage. These include having on hand at all times:

- Water—one gallon per person, per day (3-day supply for evacuation, 2-week supply for home stays).
- Food—non-perishable, easy-to-prepare items (3-day supply for evacuation, 2-week supply for home stays).
- Flashlight (Do not use candles during a power outage due to the extreme risk of fire.).
- Battery-powered or hand-crank radio (NOAA Weather Radio, if possible).

ASSEMBLING A SOLAR GENERATOR

- Extra batteries.
- First aid kit.
- Medications (7-day supply) and required medical items.
- Multi-purpose tool (screwdriver, knife, pliers, etc).
- Sanitation and personal hygiene items.
- Copies of personal documents (medication list and pertinent medical information, deed/lease to home, birth certificates, insurance policies).
- Cell phone with charger.
- Family and emergency contact information.
- Extra cash.
- If someone in your home is dependent on electric-powered, life-sustaining equipment, remember to include backup power in your evacuation plan.
- Keep a non-cordless telephone in your home. It is likely to work even when the power is out.
- Keep your car's gas tank full.

Of course, few of us are ever fully prepared during an emergency. Increasingly, people are turning to gasoline or diesel-powered generators to power their way through periodic outages. However, traditional gasoline and/or diesel generators present a number of problems and concerns.

Each year dozens of people die from using portable fossil fuel powered generators. For example, nine deaths were blamed on carbon monoxide poisoning from the use of generators during Hurricane Sandy (*Huffington Post, 11-02-2012*).

CHAPTER 2: IN CASE OF EMERGENCY

In addition to carbon monoxide poisoning, fires caused by pouring gasoline or diesel fuel into hot generators is a constant threat.

Traditional generators are very loud, a source of noise irritation in urban or suburban neighborhoods. They are also smelly, and in tightly clustered neighborhoods the fumes can permeate throughout the area.

When the electricity is down, so too are the gas pumps. Fuel may not be available during an extended power outage.

Solar generators avoid most of these problems. However, it should be noted that solar generators produce electricity – which when mishandled can still cause injury and/or death. So care should be exercised, even with solar generators.

8  Assembling a Solar Generator

Chapter Three
Energy efficiency

As the cost of electricity soars, the value of taking simple energy efficiency steps in your home makes more and more sense – every day – not just during an emergency. By installing energy-efficient appliances, **LED lighting** systems, and weather-proofing your home, you will not only save money year-round, but will find that you can get by much more easily with a smaller generator during a power outage.

Energy efficiency measures include:
- replacing light bulbs with **compact florescent** or LED bulbs.
- using **Energy-Star** appliances when possible.
- hooking up electric devices and microwaves (anything that displays an indicator light when not in use) to power bars and turn off the power bar when the device is not in use.
- adding insulation to walls and attic.
- caulking or replacing leaky windows.
- hanging your clothes to dry on a line rather than using an electric dryer (whenever possible).

- turning down the temperature setting on your thermostat.
- turning off lights when not in use.
- insulating your hot water heater.
- insulating hot water pipes.
- fixing any leaking heating duct work.
- using smaller appliances when cooking (a toaster oven versus a full size oven, for example).
- replacing filters (on furnaces, refrigerators, etc) often.

Chapter Four
Generator Parts

Now that you've decided you want to design and build your own solar generator, you will need to identify all the major parts of the system and understand how they function.

A solar generator is basically a very small **stand-alone photovoltaic system**. As you learn about and build your generator, you are really practicing skills in designing and building one of the most complex PV systems on the market. The only difference between your solar generator and a system for a cabin in the woods is a matter of scale.

With this in mind, let's first take a look at the various components of a typical stand-alone PV system. As demonstrated in Figure 4-1, these components include:
- the solar array (or solar panel)
- a charge controller
- the battery bank
- an inverter
- some disconnects
- load distribution (getting power to appliances)

Assembling a Solar Generator

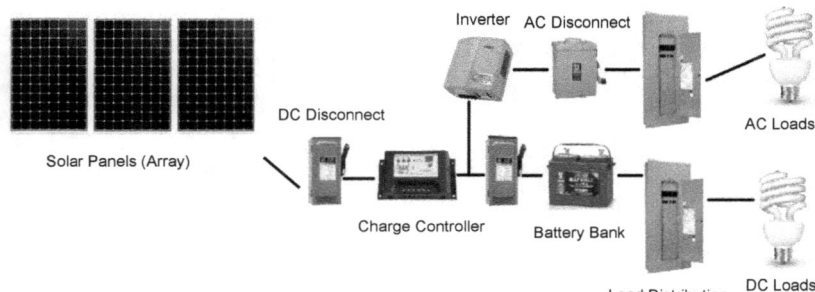

Figure 4-1: Typical Stand-Alone PV System

Solar Panels:
The **solar panel** (or **PV module**) is the energy source of the system. When exposed to the sun, solar panels will generate electricity that can then be stored in the battery. The panels (more than one panel connected together is referred to as an **array**) are rated in **nominal volts,** as well as in watts. These ratings will be important later when we design and build the system. The larger the solar panel or array, the faster the battery bank will recharge after use. The power generated by a solar panel is in the form of DC, or **direct current** power. This is the same form of power stored in batteries.

DC Disconnects:
It is important to have a way to disconnect every power source from the rest of the system. This disconnect can be in the form of a switch, or for smaller systems, it can be accomplished simply by unplugging the array and/or the battery bank.

Charge Controller:
The **charge controller** monitors both the solar array (to see if there is energy available) and the battery bank (to see if energy is needed) and controls how the battery is charged by the solar array (hence the name).

CHAPTER 4: GENERATOR PARTS

Battery Bank:
The function of the **battery bank** is to store the DC energy from the solar panels until it is needed, and then provide that power to service the loads (light the lights, run the refrigerator, etc). PV systems (both in the "real world" and in our solar generator) use **deep cycle batteries**. These operate much differently than a typical car battery. Car batteries cannot be used in a PV system or in a solar generator.

Inverter:
The **inverter** converts DC power from the battery bank into AC (**alternating current**) that can be used by appliances and/or lights. The power from the utility company is in the form of AC power - so the inverter's power output will closely match what would normally be provided by the utility.

AC Loads:
Anything that would normally be plugged into a wall outlet in the home will very likely be an **AC load**. Examples include: a lamp, a television set, a freezer, water pump, etc.

DC Loads:
Any electronic device that contains a battery is likely using DC electricity rather than AC. Your mobile phone, for example, is a **DC load**. Most stand-alone PV systems allow for some items to be charged or operated directly from the battery bank (since it is already DC current), rather than converting the energy from DC to AC and then back to DC again (the little black box connected to the cord used by your laptop, for example, converts AC current from the wall outlet to DC current that can be

used to run your laptop or charge the battery in the laptop.

Your solar generator will incorporate very similar components, as shown in Figure 4-2.

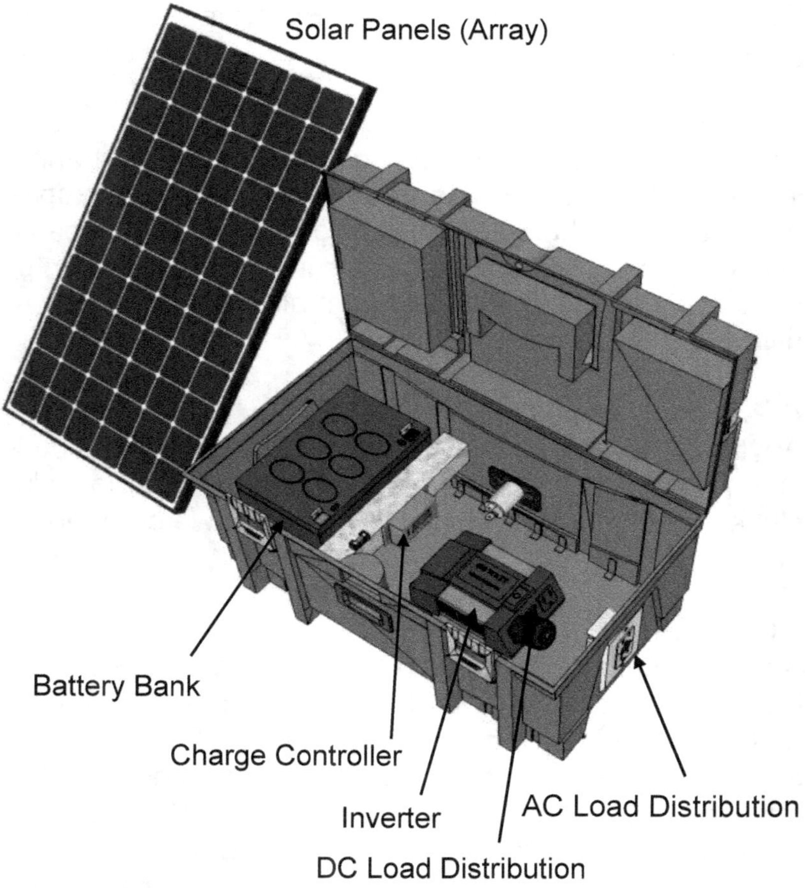

Figure 4-2: Typical Solar Generator Components

Voltage Requirements and Compatibility:

When designing a stand-alone PV system, there are a number of voltage options available. **The National Electrical Code (NEC)** limits residential battery bank voltage to less than 50 volts. So realistically (based on product availability and industry norms), most systems are designed to be either 12-volt (DC), 24-volt (DC), 36-volt (DC) or 48-volt (DC) configurations.

All the components (the solar panels, the charge controller, the battery bank and the inverter) must be compatible - able to operate at the same nominal voltage as all other components.

With this in mind, and based on product availability, you should plan on designing your solar generator to operate as a nominal 12-volt system.

This means that:
- your solar array should be configured to 12 nominal volts,
- your charge controller should be rated at 12 nominal volts,
- your battery bank configured to 12 nominal volts,
- and your inverter rated at 12 nominal volts.

Assembling a Solar Generator

CHAPTER FIVE
SELECTING THE INVERTER

Note: *Load requirements determine the size of the inverter to install in the generator.*

You will need to determine the load requirements that the generator will service. More simply, what appliances do you intend to plug into the generator and how much power do they require?

Most appliances have labels that indicate how much power they need to operate properly. You can often locate this information on a label attached to the the back or bottom of the appliance.

For example, let's assume you intend to use a food steamer with your solar generator. You should be able to find a label attached to the back of the appliance, similar to the one illustrated in Figure 5-1.

ASSEMBLING A SOLAR GENERATOR

Labels of this type will vary from manufacturer to manufacturer, however there are three very specific bits of information that you will need to find.

First (in the USA anyway, this may be different in other countries), make sure the appliance operates at 120 volts (V). Sometimes it may be noted as 115 volts and even occasional 110 volts - but it all means pretty much the same thing, since the voltage under which an appliance operates is a range, rather than a specific number.

Figure 5-1: Typical Energy Use Label

Second, you will also want to verify that it operates at 60 **hertz (Hz)**. If you have used the appliance previously by plugging it into a typical household receptacle - then you can pretty much assume that it does operate at 120 V, 60 Hz (again, in the US - other countries have different standards).

Third, you will need to discover how many watts (W) the appliance draws when it is running. In the above example, the Oster Food Steamer requires 900 watts to operate.

CHAPTER 5: SELECTING THE INVERTER

Calculating Loads (if no label available):

In some cases, the manufacturer does not provide the number of watts, but does provide the voltage and the number of amps the unit uses when operating.

In these cases you can calculate the watts simply by multiplying volts times amps to equal watts. For example, if Oster had stated that its model 5716 Food Steamer operates at 120 V and 7.5 amps, then you would calculate the watts at 900 (120 V x 7.5 amps).

It is the total load requirements that determine the size of the inverter needed within the generator. If the solar generator was designed solely to service the Oster food steamer in our example, then you would need to select an inverter that produces 120 V 60 Hz power and is rated at a size of 900 watts or larger.

If you attempt to plug the food steamer into a unit rated at less than 900 watts, the inverter will either not work or, worse yet, might blow a fuse within the generator and/or inverter.

Bear in mind that you may plug more than one appliance into the generator at any one time - so it is the TOTAL power plugged into the generator at any one time that is important.

So, for example, if you decided to also plug in a computer that draws 300 watts of power into the solar generator at the same time that the unit is powering the food steamer, the system will now be servicing 1,200 watts.

If the inverter had been sized at 1,000 watts, for example, then the food steamer will work fine, but once the computer is added to the circuit, it will overload the capacity of the generator and the unit will shut down and/or blow a fuse.

If you are unsure of the power draw of a specific appliance, there are devices, such as the **Kill A Watt meter** illustrated in Figure 5-2, that can help in determining the load of a specific appliance.

Simply plug the meter into an AC power outlet, then plug the appliance into the meter. Set the meter's indicator to display watts - and the unit will display how many watts are being consumed by the appliance.

This meter is also capable of displaying the volts and hertz available at the outlet, as well as the amps and watt-hours consumed by the appliance connected to the meter.

Figure 5-2: Kill A Watt Meter (Photo by P3 International)

Some appliances (such as a refrigerator and/or a freezer) only draw electricity periodically when in operation. For example, a refrigerator may only actually be cooling (using electricity) ten minutes out of every hour. The best way to determine the actual power draw of the appliance is to plug the unit into your Kill A Watt meter, set it to the KWH setting, and leave it plugged in for a period of time

Chapter 5: Selecting the Inverter

(say 24 hours). You will then know how much power the unit uses over time.

It will become apparent when testing the various appliances around your home, that those appliances that convert electricity into heat require a very large amount of power to operate. So much power, in fact, that it may prove impractical to operate them with a solar generator.

For example, a standard toaster (as seen in Figure 5-3) will draw more power than a full-size freezer, a lamp, a laptop computer, a printer, a phone charger, and a clock radio - combined. So you may find it much more practical to forego toast during you next power outage, and in doing so, you can get by with a smaller generator.

Table 5-1: Appliances that typically draw less than 400 Watts

Appliance	Continuous Watts
Laptop Computer	45 watts
Energy Star Printer	10 watts
Box Fan	75 watts
Lamp w/ Compact Florescent Bulb	14 watts
Cordless Telephone	2 watts
Television	40-60 watts
DVD Player	5 watts
Standard Refrigerator	200-350 watts
Food Processor	200 watts
Clock Radio	70 watts

 ASSEMBLING A SOLAR GENERATOR

Most electronic devices, such as smart phones, gaming consoles, and MP3 players draw so little power (less than 5 watts) that they hardly tax even the smallest of solar generators.

Figure 5-3: A toaster drawing 841 watts when running

Loads with Motors:

Some appliances with motors, such as water pumps or refrigerators, draw more electricity when they start up than they require when running. You will have likely noticed this effect in your home when an appliance (such as a pump) briefly dims the lights upon startup.

The amount of power drawn at startup may be as high as 300% of the power required to run the appliance after it is up and running.

In the world of inverters, the power required to run the appliance for the long haul is referred to as its **continuous capacity.** The higher power required for a few seconds when the appliance starts up is referred to as **surge capacity**.

Chapter 5: Selecting the Inverter

Typical inverters have a surge rating twice that of their continuous capacity. Generally this capacity should be more than enough to handle the momentary power requirements upon startup.

However, don't necessarily believe the surge capacity ratings of small and inexpensive inverters. Our experience has shown that they are often, shall we say, "optimistic." You will be much better served selecting an inverter with a much higher capacity than required if you are going to service loads with motors.

Table 5-2: Appliances that Incorporate Motors:

Appliance	Continuous Watts
Deep Freezer	400 watts
Standard Refrigerator	200-350 watts
Washing Machine	500 watts
Hair Dryer	600 watts
½ horsepower pump/motor	373 watts
1 horsepower pump/motor	746 watts
Blower fan on Gas Furnace	1000 watts

Loads that Generate Heat:

Appliances that convert electricity into heat draw a great deal of power. Loads of this type might be practical to operate with only the largest of solar generators.

One exception may be using the solar generator to operate the blower fan on a gas furnace. The heat of a gas furnace is generated by burning natural gas. The fan simply distributes this heat throughout the building.

Table 5-3: Appliances that Generate Heat

Appliance	Continuous Watts
Portable Space Heater	1400 watts
Coffee Maker	1200 watts
Electric Water Heater	2,000-5,000 watts
Hair Dryer	600 watts
Toaster	1000 watts
Toaster Oven	1000 watts
Two-Burner Hot Plate	1600 watts
Clothes Dryer	5,000 watts
Electric Range	5,000-8,000 watts

Other Factors to Consider when Selecting an Inverter:

The primary issues to factor in when selecting an inverter include:

- The inverter's output matches the nominal voltage of the appliance.
- The inverter's can accept the input nominal voltage of the battery bank.
- It matches the Hertz (wave cycles per second) of the appliance.
- It has a continuous power rating large enough to handle all the appliances plugged into the generator.

One other issue to consider is the **waveform** of the electricity generated by the inverter.

CHAPTER 5: SELECTING THE INVERTER

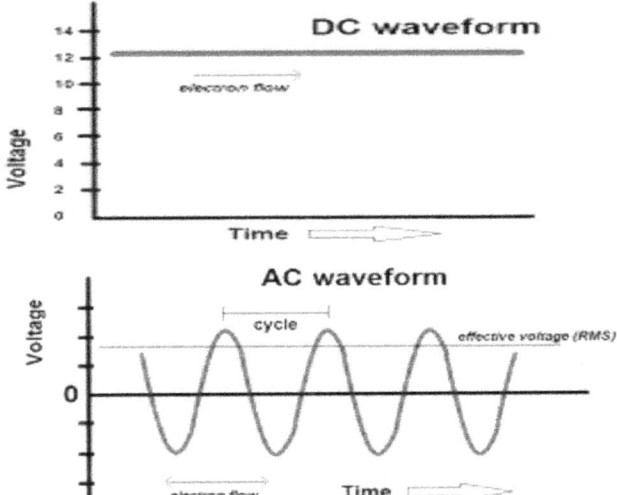

Figure 5-4: DC and AC electrical signal waveforms

The principal role of the inverter is to transform the direct current (DC) waveform that comes from the battery into an alternating current (AC) waveform that can be used by the appliance.

The waveforms of DC and AC power are quite different, as shown in Figure 5-4. The inverter must convert the flat line wave of DC power into a form that more closely resembles the wave form of AC power.

Early inverters managed to convert the waveform in a digital, on-off process. This was referred to as a **square wave inverter**. These low-cost inverters often did not work with sensitive electronics and/or with motors. This is an old technology and there are very few of these inverters on the market today.

As the technology matured, inverters changed the conversion process, more closely resembling the waveform

ASSEMBLING A SOLAR GENERATOR

Figure 5-5: Square wave, modified sine wave and sine waves

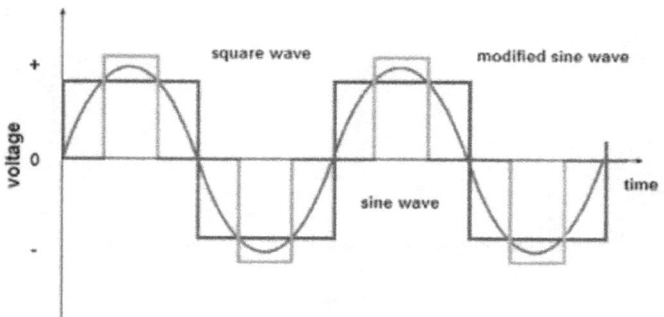

of AC power. These **modified sine wave inverters** are inexpensive and generally work well. They may, however, cause a bit of interference (a buzzing noise) in some sensitive electronics.

The best inverter is a **pure sine wave inverter,** that matches exactly the waveform provided by the utility company to your home. There should be no compatibility issues if you use a pure sine wave inverter. The three wave forms are illustrated in Figure 5-5.

Some appliances, such as motors and microwave ovens, will only produce full output with sine wave power. A few appliances, such as bread makers, light dimmers, and some battery chargers require a sine wave to work at all.

Inverter Safety Issues:

1) When operating, the inverter typically is cooled by a fan. Make sure to vent your solar generator to allow for the exchange of air. Also, ensure the fan is clear and free of obstructions.

Chapter 5: Selecting the Inverter

2) Connect only to a power source of a similar rating as the inverter (12-volt, for example). Higher voltage may damage the unit.

3) Make very sure the polarity of the input connections are not reversed. Connect the negative terminal of the battery bank to the negative input terminal on the inverter. Connect the positive terminal of the battery bank to the positive input terminal on the inverter. Reversing the polarity may damage the inverter.

4) Keep the inverter dry and clean.

5) Connect the positive terminal from the battery to the inverter last (the system is grounded through the negative terminal).

 ASSEMBLING A SOLAR GENERATOR

CHAPTER SIX
THE BATTERY BANK

Note: *Determine how much energy storage you need. It is the battery bank that provides the power - so the larger the battery bank, the more power will be available for a longer period of time.*

In a stand-alone photovoltaic system (or a solar generator), it is the batteries that power the loads (not the solar panels). For the most part, the solar panels serve only to recharge the batteries.

So the larger the battery (or **battery bank** if there is more than one battery), the longer the load can be powered before the battery bank must be recharged.

Remember, the battery bank does not determine the size of the load that can be serviced. It is the inverter that determines the size of the load the solar generator can power. The size of the battery bank will simply determine how long that appliance can run on the power stored in the solar generator.

ASSEMBLING A SOLAR GENERATOR

To fully understand how this works, we must once again refer to the **power equation** of Watts = Amps x Volts.

This equation is helpful when addressing "instantaeous" power demands. For example: When sizing the inverter for your solar generator, if you know that an appliance, when running, requires 4 amps of current, and runs on 120 volts, then it is drawing 480 watts of power and you will need to use an inverter that can handle that continuous load demand.

This doesn't, however, address how long you can run the appliance. We know it needs 480 watts of power - but will it run for one second, or one hour, or several days? We need to incorporate time into our equation.

With this in mind, we will modify the power equation to reflect watt-hours. This measurement is how we consume electricity. When we buy power from the electric company, we purchase it in quantities of kilowatt-hours (1,000 watt-hours).

A watt-hour is simpy one watt of power consumed at a constant rate for a period of one hour.

When you receive your electric bill, it notes how many kilowatt-hours (1,000 watt-hours) your household consumed over the period of a month. Unlike the electric grid, which is assumed to have a limitless supply of power (until it doesn't, as in a power outage), your solar generator has a limited supply of power available. This available power is stored in the battery bank.

CHAPTER 6: THE BATTERY BANK

So our new power equation, modified to incorporate time, now reads:

Watt-Hours (Wh) = Amp-Hours (Ah) x Volts (V)

Note that voltage remains unchanged, but amps are now modified to deal with the amount of time the electrical current is available.

Fortunately for us, at least in the design and construction of solar generators, batteries are rated (in capacity) in **amp-hours (Ah).** An 8-Ah battery is smaller than a 20-Ah battery which is smaller than a 50-Ah battery... and so on.

In theory (although we will see it is not quite this simple), a 12-Ah battery can supply one amp of current to a load (operating at the voltage of the battery) for a period of 12 hours. A 50-Ah battery can supply that same current for a period of 50 hours, etc.

However, deep cycle batteries, the type used in solar generators, do not use all the power available. They do

Figure 6-1:

Typical "small" deep cycle battery. This is a 12 volt, 7 amp-hour sealed glass matt lead acid battery.

not drain down to zero. They will always have some reserve in them when the system considers them "empty". So you will not, in reality, get 12 hours of one-amp current from a 12-Ah battery. It will be something less than this (this will be regulated by the inverter or the charge controller used in the system).

Also, the more quickly the electricity is used, the more quickly the battery will be drained. In other words, if you only use only a little power - the battery may actually provide current for much longer than the amp-hour rating might suggest. If you use a lot of power (draining the battery quickly), it may provide much less power than its rating indicates.

As a result, battery manufacturers have standardized on measuring their batteries at a discharge rate draining from full to empty over a 20-hour period of time. This **C-20 rating** is a way of comparing "apples to apples."

Figuring out how long a battery will supply power to a load is not as straight forward as you might expect. So we must be content with knowing that a 12-Ah battery will

Figure 6-2:

Typical "mid-sized" deep cycle battery. This is a 12 volt, 91 amp-hour sealed liquid lead acid battery.

(Image from Trojan Battery Company)

Chapter 6: The Battery Bank

supply less power than a 25-Ah battery, and leave it at that until the unit is tested under "real world" conditions.

Depth of Discharge:

Not all of the power that is contained in the battery is available to be used. A battery that has 10% of the power remaining is said to be at 90% **depth of discharge (DOD)**. A battery with 40% of the power remaining is at 60% DOD.

As a battery discharges, from full to empty, the voltage contained within the battery actually drops. A 12-volt battery rarely will give a reading of exactly 12 volts when tested with a volt meter. Referring to it as a 12-volt battery is simply a shorthand method of identifying its voltage range, by referring to the battery's **nominal voltage**.

Table 6-1: Voltages at various Depth of Discharge:

State of Charge	Depth of Discharge	12-Volt Battery
100%	0 %	12.73 volts
90%	10%	12.62 volts
80%	20%	12.5 volts
70%	30%	12.37 volts
60%	40%	12.24 volts
50%	50%	12.12 volts
40%	60%	11.96 volts
30%	70%	11.75 volts
20%	80%	11.58 volts
10%	90%	11.31 volts
0%	100%	10.5 volts

ASSEMBLING A SOLAR GENERATOR

As indicated on Table 6-1, a 12-volt battery measuring 12.1 volts (on a volt meter) is actually half empty, or at a 50% depth of discharge.

System Voltage Range:

The inverter you select will likely have an operational range. If the input exeeds a certain voltage, it will not operate. Likewise, when the voltage within the battery drops below a preset minimum, the inverter will (typically) sound an alarm and then shut down if the voltage drops further.

For example, an inverter might have an operational range of 11.5 volts to 15.3 volts.. If the input voltage is higher than 15.3 volts (such would be the case if it was hooked up to a 24-volt battery bank), then the inverter would not operate. When the voltage drops below 11.5 volts (a situation where the 12-volt battery would still have about 15% of its capacity available) - it will likewise shut down.

A warning alarm may sound when the battery still has 20% of its capacity remaining - letting you know that the battery needs to be recharged for continued use.

Note: Loads with Motors
The power surge at startup from a load with a motor may pull the voltage level in the battery down dramatically - too low, perhaps, for the inverter to function.

As a result, loads with motors may require that the level of charge within the battery be higher than otherwise might be the case. For example, you may only be able to use half the charge of the battery (50% depth of

Chapter 6: The Battery Bank

discharge) rather than 90% of the battery's power. In this way the generator will have enough power available to handle the surge.

Some larger and more expensive inverters may allow you to adjust the minimum voltage. However in most cases you will simply have to allow for a larger battery bank when operating loads with motors.

Sizing the Battery (or Battery Bank):

In constructing your solar generator, batteries are perhaps the most vexing problem you will face. They are expensive and heavy. So the more power available in the unit, the heavier it will be and the more expensive it will be to build.

But the point of having a solar generator is to ensure that power is available when needed. So you will need to do some calculations in order to strike the right balance (cost/weight versus power availability).

For example:
You work from home and must have your laptop computer, printer, modem and cordless telephone available at all times.

You have tested these appliances and find that the laptop computer draws 60 watts, the printer draws 10 watts, the modem pulls only 2 watts, and the cordless phone requires 3 watts. You wish to make a solar generator that allows you to plug all these items in and use them all at the same time. You also want to make sure you can use them for 8 hours a day, for as long as 3 days (three

 ASSEMBLING A SOLAR GENERATOR

miserable days with no sun available to recharge the batteries).

First, calculate the total watts that will be used at any one time. In this case it will be: 60 watts (computer) + 10 watts (printer) + 2 watts (modem) +3 watts (cordless telephone), for a total of 75 watts. So the inverter you install must be rated to handle at least 75 watts, although a larger inverter would be fine as well (and possibly not much more expensive).

Now you must factor in time. 8 (hours per day) x 3 (days) = 24 hours of availability required.

So the number or watt-hours can be calculated simply by multiplying watts (75 in this case) by hours (24 in our example) to equal 1,800 watt-hours (Wh).

Assuming the selected battery produces 12 volts, and that the inverter will allow us only to use 80% of the battery's capacity (80% DOD) - we have all the data we need to determine the size of the battery required.

Returning to our modified power equation, we know that:

Watt-Hours = Amp-Hours x Volts

Plugging in the numbers...

1,800 Wh = Ah x 12 V
Ah = 1,8000 Wh / 12 V = 150 Ah

So to power our workstation for 24 working hours, we will need 150 amp-hours of stored energy. But we can't just go buy a 150 Ah battery, since we can only use 80% of

Chapter 6: The Battery Bank

the power contained in the battery. So another calculation is required.

Battery Size Required = 150 Ah / 80%
15 Ah / .80 = 187.5 Ah

Almost finished. One more factor must also be taken into consideration. The inverter in the generator will convert the DC power from the battery into usable AC power. But this process is not 100% efficient. Some power is lost in the conversion process. With small inexpensive inverters, usually about 10% of the DC power is lost. We need to factor in this loss.

Adjusting for the inverter inefficiency results in:

Battery Size Required = 187.5 Ah / .90 (inverter inefficiency)
Battery Size Required = 208.3

So we now know that we need a battery rated at 208.3 Ah or more to service the load we anticipate. Although a 200 Ah battery might do the trick, since our estimate of time required was fairly arbitrary.

Creating a Battery Bank:

For smaller solar generators, typically one battery will meet the needs of the system. But for larger systems, it may be necessary to hook more than one battery together to provide enough stored energy to meet the load demands.

In order to do this, it is necessary to understand a couple more electrical terms.

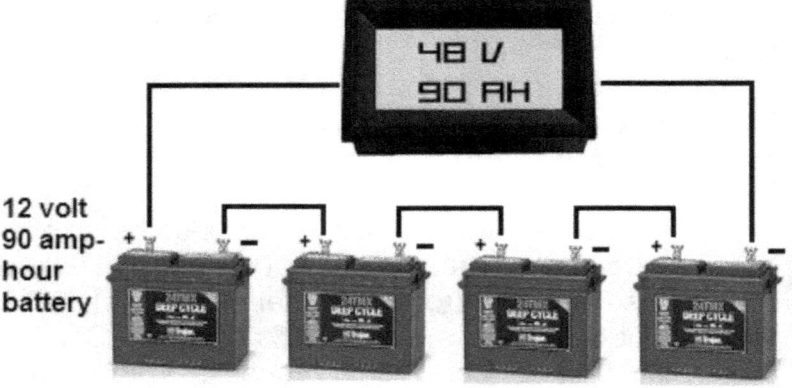

Figure 6-3: Four 12-volt, 90 amp-hour batteries connected together in series.

When the positive terminal of one battery is connected to the negative terrminal of another battery (and the other two terminals typically are connected to the inverter), these batteries are said to be connected in **series**.

When two batteries (or solar panels for that matter) are connected together in series, the voltage increases but

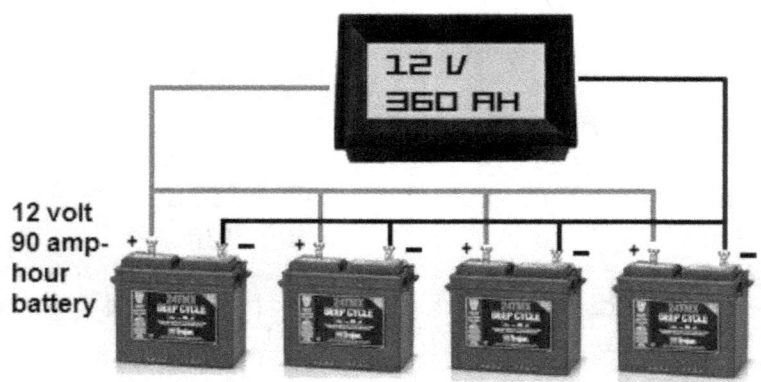

Figure 6-4: Four 12-volt, 90 amp-hour batteries connected together in parallel.

CHAPTER 6: THE BATTERY BANK

the amp-hours stored within the battery bank remains the same as a single battery within the system.

For example, if you connect four 12-volt, 90 amp-hour batteries together, as indicated in Figure 6-3, the nominal voltage of the battery bank will increase to 48 volts, but the storage capacity of the battery bank will remain at 90 amp-hours.

When the positive terminals of two or more batteries are connected together, and the negative terminals are connected together - these batteries are said to be connected in **parallel**. When connected in this way, as shown in Figure 5-4, the voltage of the system remains the same as a single battery within the system - but the amp-hours or storage capacity increases.

In this example, the overall system voltage (when connected in parallel) remains at 12 nominal volts, but the amp-hour storage capacity of the battery bank is increased four fold, to 360 amp-hours.

Typically there will be two reasons for incorporating more than one battery in a solar generator. First, the battery selected may only be rated at a nominal 6 volts. In this case, two of these batteries connected in series will be necessary in order for the battery bank to be compatible with the entire 12-volt solar generator system.

The second, more common reason, is to increase the storage capacity of the generator. When additional 12-volt batteries are added to the battery bank for this reason, they will always be connected together in parallel.

Other Issues When Selecting a Battery:

Far and away the most common type of deep cycle battery is the lead acid battery. There are other types as well, however, such as the **lithium ion battery** and the **nickel cadmium battery** - but for the sake of this text, we will assume these technologies are out of the price range for a typical do-it-yourself solar generator.

But even within the world of lead acid batteries, there are choices.

Sealed versus Unsealed Batteries:
If money were not an issue, **sealed batteries** would be the battery of choice. There is less chance of acid spilling (should the generator be dropped or knocked over) and a slightly less chance of out-gassing should the battery be overcharged for some reason. But sealed batteries are typically more expensive than similarly sized unsealed alternatives.

Liquid, Gel or Glass Matt:
The sulfuric acid in the battery can be stored in the form of a liquid, a **gel**, or a fiberglass mat-like material (**glass matt**) soaked in acid. The advantage of the gel or glass matt over the liquid is that if the battery is damaged, it is less likely that acid will leak from the battery. Again, however, gel and glass matt batteries are typically a bit more expensive than liquid lead acid batteries.

Safety Concerns When Working With Batteries:

1. Lead acid batteries contain sulfuric acid. This acid can cause serious injury if it touches your skin or eyes, or is inhaled or swallowed.

2. Use ANSI*-approved safety goggles and electrically insulated gloves while working near batteries. Should acid come in contact with your skin or eyes, flush with clean water. The acid can also be neutralized with baking soda.

3. Locate batteries in a clean, well-ventilated area, away from any ignition sources and flammable materials. Lead-acid batteries (even sealed batteries) release small amounts of explosive hydrogen gas while charging. Make sure to properly vent your solar generator to avoid the build-up of any hydrogen gas.

4. Only connect similar batteries together. Do not mix gel, glass matt and liquid (wet) batteries in the same battery bank. Also, do not mix old and new batteries. If you must replace one battery, it is best to replace them all - since the entire bank will operate at the level of the poorest performing battery.

5. Store your generator in a cool environment when not in use. The battery bank in your solar generator will last longer if you store it in a cool, well ventilated location.

6. Be sure to install a properly rated fuse to the positive output terminal of the battery bank in the wiring system leading to the inverter (more on this later).

*ANSI refers to the American National Standards Institute

 ASSEMBLING A SOLAR GENERATOR

Chapter Seven
The Solar Panels

Note: *When sizing the solar panel or array, the larger the panel, the faster the battery bank will recharge.*

A solar generator just wouldn't be a solar generator without solar panels to provide the power for the unit.

The price of solar panels has come down dramatically in recent years, but the solar panel will still be one of the more costly components within the solar generator. The larger the solar panel, the higher the cost.

For large arrays, it is now possible (February, 2015) to purchase quantities of relatively large panels for somewhere between $0.60 to $1.20 per watt. So a 250-watt solar panel will retail for somewhere between $150 - $300. But when purchasing a smaller panel (and only one at a time, so no discount for quantity), the price per watt will be considerably higher.

A small 10-watt solar panel can be purchased (with some searching) for around $4 per watt (around $40). Larger 50-watt panels will run about $2 per watt (or about $100).

As becomes evident, the larger the panel (or array), the more expensive the project. However, the larger the panel (or array), the faster the battery bank will recharge. So constructing your solar array will be a tradeoff between time and money.

Most people intend to use their solar generator infrequently, only duirng times of emergency. So hopefully these emergencies occur with a considerable time lag between each event. As a result, you may be able to get by with a much smaller solar panel than you might think.

Selecting the Panel:

Solar panels typically come on one of three forms. These include **monocrystaline** (each solar cell made from a thin slice of a single silicon crystal illustrated in Figure 7-1), **polycrystaline** (each solar cell made from a thin slice of a block of molten crystals shown in Figure 7-2), or **thin film** (also called **amorphus**, made from a flexible material shown in Figure 7-3).

Figure 7-1:

Typical monocrystaline panel (note the individual cells divided by a diamond shape)

CHAPTER 7: THE SOLAR PANELS  45

Figure 7-2:

Typical polycrystaline panel (note the bits of crystal visible throughout, and normally the panel is bluish in color)

It really doesn't matter which of the three you select, as long as you simply refer to the watts each is rated to produce. A monocrystaline panel is typically more efficient than a polycrystaline panel which is more efficient than a thin film panel. But if each panel is rated to produce 10 watts, then each will produce 10 watts of power when exposed to full sunlight. The monocrystaline panel may be a bit smaller than a similarly rated thin film panel (because it is more efficient), but if size is not an

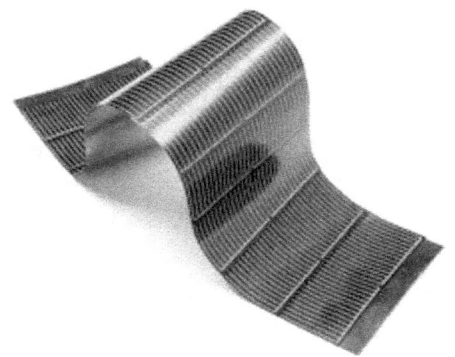

Figure 7-3:

Typical thin film or amorphus panel (while the panel is flexible, they are often mounted in glass - which creates a rigid solar panel).

(Image from sunconnect.com.au)

issue - then the power output from all should be the same.

Assuming your generator will be a 12-volt system, you will need to select a solar panel that is rated as a nominal 12-volt panel (or two 6-volt panels connected together in series).

Standard Test Conditions:

In order to allow an "apples-to-apples" comparison, all solar panels are rated by the amount of energy they produce under **standard test conditions**. It almost doesn't matter what these conditions are, as long as all the manufacturers agree and rate their products in a similar manner.

However, in the world of solar panels, standard test conditions dictate that the panel has been rated at:

- sunlight of 1000 watts per square meter (this is the power available in full sunlight, without clouds),
- a cell temperature of 25° C (78° F),
- and atmospheric pressure of 1.5 (basically at sea level).

As you gain experience with the solar generator, it will become clear that the panels will recharge the battery much faster in full sunlight (even on a cloudy day they will produce some energy - but not nearly as much as sunny days) and actually perform better in cold weather than in hot.

But the main factor in how quickly the batteries are recharged will depend on the size of the panel and the

CHAPTER 7: THE SOLAR PANELS 47

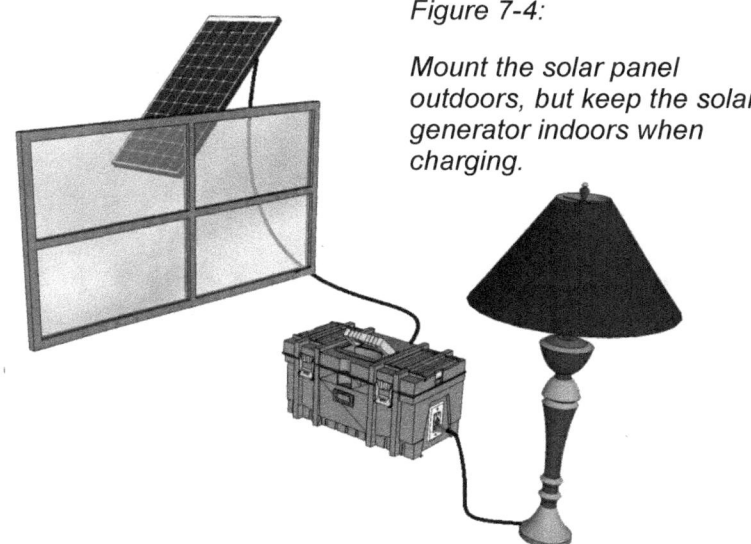

Figure 7-4:

Mount the solar panel outdoors, but keep the solar generator indoors when charging.

amount of direct sunlight hitting the panel for the longest amount of time possible.

Where to Place the Panel:

To get the most sunlight, the solar panel will need to be placed outdoors. On the other hand, you will want to store your battery (as well as the generator itself) in a cool, dry location.

So the ideal situation will be to mount the panel outdoors, but store the generator indoors (connecting the two together with a properly designed power cable (more on that later) as illustrated in Figure 7-4.

The solar panel is designed to withstand the moisture and other harsh conditions that it will experience outdoors.

ASSEMBLING A SOLAR GENERATOR

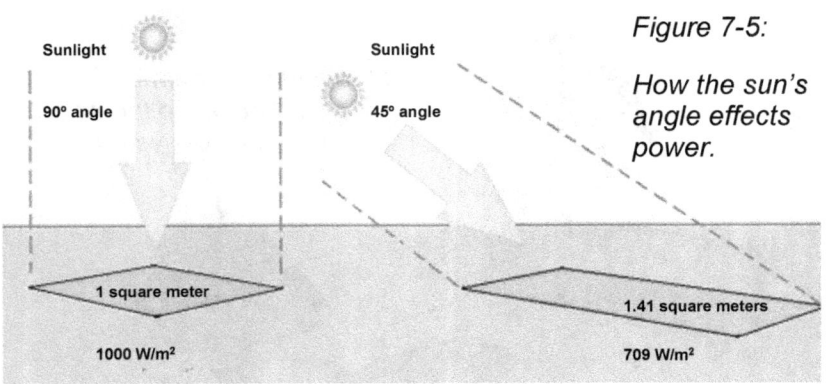

Figure 7-5: How the sun's angle effects power.

The solar generator may not be so robust, plus the batteries may be damaged if stored in the sunlight or allowed to freeze.

You may be tempted to simply place the panel in sunlight coming indoors through a window, but often windows incorporate UV (ultra-violet light) protection, which can greatly reduce the amount of energy in the available sunlight. So place the panel outdoors if at all possible.

Positioning Your Solar Panel:

The angle at which you place the solar panel will greatly impact how much energy it will gather from the sun. The ideal situation would be to have the face of the solar panel directly facing the sun at all times. This ideal will likely be impractical, as it would require a system that reorients the panel's position throughout the day and throughout the year.

As demonstrated in Figure 7-5, adjusting the angle of the panel from 90 degrees (facing directly at the sun) to 45 degrees will reduce the amount of energy striking the panel by nearly 30%.

While it may be impractical to adjust the angle of the panel during the day as the sun tracks from east to west (the **azimuth**), you may wish to design a mount so you can adjust the angle throughout the year (adjusting for **altitude**) as demonstrated in Figure 7-6.

If you happen to live in the northern hemisphere, you will want your panel to face as close to **solar south** as possible (assuming there is no tracking system that

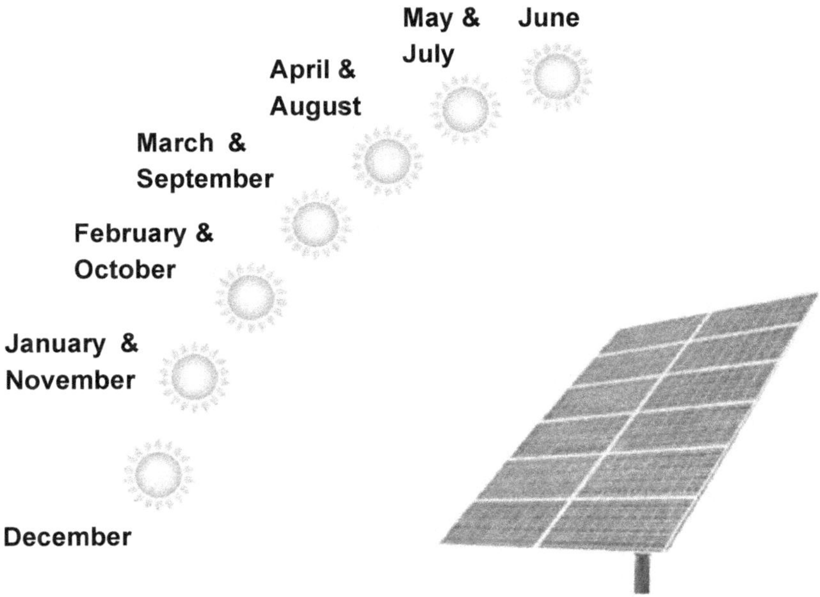

Figure 7-6: You will greatly increase the power gathered by your solar panel if the mount is designed so that the altitude angle can be adjusted throughout the year.

adjusts the panel for azimuth as the sun tracks across the sky during the day).

It is important to note that solar south and **magnetic south** are not the same thing. In some parts of the world they will vary by a significant amount (as much as 20 degrees). So you cannot rely on a compass to orient your panels.

Fortunately technology has come the rescue. Most people who own a smart phone can upload a simple app that will help orient the panel to **true south** (or solar south). You can also use an app on the smart phone to help set the angle of altitude of the panel, making calculations unnecessary.

What Size Solar Panel do you Need?

As mentioned earlier, the role of the solar panel in this system is to recharge the battery bank. The larger the panel (as rated in watts), the faster it will recharge the battery.

Available sunlight varies widely depending on where you are located. If you live in Phoenix, AZ, you have (on average) about twice as much sunlight each day as you would have available if you live in Seattle, WA.

In an effort to compare and measure available sunlight across the nation, the solar industry refers to available sunlight (**irradiance**) in terms of **peak sun hours**.

This means that if a square meter of land in a particular place receives 5,000 watts of energy (from the sunlight) over the course of a day, that location will be said to

CHAPTER 7: THE SOLAR PANELS

experience 5 hours of peak sunlight (sometimes called **insolation**).

Clearly the power within the sun varies throughout the day, as clouds pass and the angle of the sun changes.

So referring to a location with 5 hours of insolation does not mean that there are 5 hours of sunshine and 19 hours of total darkness. It simply means that there will be 5,000 watts of energy from the sun per square meter available at that location on average each day.

If you wish to determine the exact amount of sunlight available at your location, the **National Renewable Energy Lab (NREL)** has a great tool called **PVWatts,** that will calculate the number of peak sun hours available at various locations when the panels are set at various angles (simply search for "PVWatts," since specific websites change from time to time).

For purposes of our calculations, we will assume the site has 4 hours of peak sun (a fairly consistent average across the USA).

By converting available sunlight into hours, we can now calculate watt-hours, or energy used over time.

For example:
Let's assume we have a 10-watt solar panel and intend to recharge a 20-Ah battery. We also assume that our location receives 4 hours of peak sunlight (on average) per day.

Daily power from the panel = 10 watts (size of the panel) x 4 hours (peak sunlight) = 40 watt-hours

If our battery and panel operate at 12 nominal volts, and watts = amps x volts, then...

40 Wh / 12 volts = 3.33 Ah

So, if our 20-Ah battery is at 80% DOD (only 20% of its charge remaining) when we began recharging it, it will take...

Days to fully recharge = (20 Ah x .80) / 3.33 Ah per day = 16 Ah / 3.33 = 4.8 days.

So it will take (on average) about 5 days to fully recharge the battery used in this example with a 10-watt solar panel.

If we had decided to connect the 20-Ah battery to a 50-watt panel instead, the battery would be recharged in just one day.

(20 Ah x .80) / (50 W x 4 hr / 12 V) = 16 / 16.66 = 0.96 days

But, assuming power outages don't occur more frequently than every week or so, in this example a 10-watt panel may be more than adequate for the task.

Larger battery banks will very likely require that they be connected to a larger solar panel.

For example:
If you wish to build a solar generator that has a battery bank with 225-Ah of storage capacity, connecting it to a 10-watt solar panel many not be adequate. With four hours of peak sunlight available each day, it will take over two months to fully recharge the system.

Chapter 7: The Solar Panels

40 Wh per day / 12 volts = 3.33 Ah per day
225 Ah battery bank / 3.33 Ah per day = 67.56 days

Incorporating a larger 50-watt panel in this system will reduce the time to recharge the battery bank to less than two weeks.

200 Wh per day / 12 volts = 16.66 Ah per day
225 Ah battery bank / 16.66 Ah per day = 13.5 days

In the summertime (when there is typically more light available during each day than in the winter), you may find the system recharges even faster.

 ASSEMBLING A SOLAR GENERATOR

Chapter Eight
The Charge Controller

Note: *The charge controller controls how the batteries are charged by the solar panels. Make sure the charge controller's nominal voltage rating is compatible with the overall system voltage.*

Charge controllers are essentially little magic boxes that monitor the state of charge of the battery bank and direct power from the solar array when needed to recharge the battery.

Since we are not designing or building charge controllers, we don't need to go into great detail as to how they work. But it is necessary to touch on a few aspects of how they function and how they are integrated into the system.

 Assembling a Solar Generator

Types of Charge Controllers:

The two major types of charge controllers that you may run across when selecting one for your system are **series** and **pulse width modulation (PWM)**.

The series charge controller (as illustrated in Figure 8-1) is an older technology and essentially works like a thermostat.

When the battery gets below a certain preset level (say for example, 13 volts or essentially anything less than 100% capacity on a 12-volt battery), the charge controller directs all the power from the solar panels into the battery. When the state of charge reaches a second preset level (for example, 14.2 volts) the charge controller will "shut off" so that the battery does not overcharge and potentially outgas or damage the battery.

Figure 8-1:

Small 12-volt series charge controller

(Image from Sunforce Products Inc.)

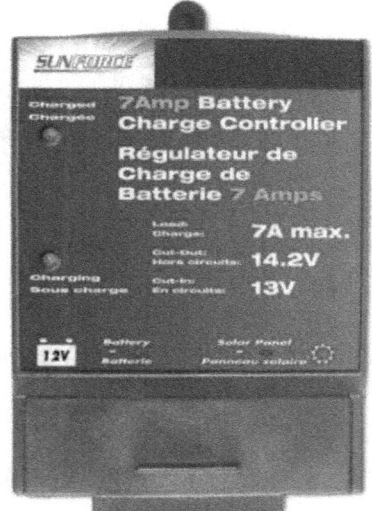

Chapter 8: The Charge Controller

Most very inexpensive charge controllers will be series type controllers. These actually work quite well for a solar generator application.

The second type of charge controller is a pulse width modulation controller. These are more sophisticated and work well to more efficiently charge large battery banks for PV stand-alone systems. Rather than an "on-off" approach to charging, they regulate the speed at which the battery is charged and slow the process as the battery approaches its capacity.

This process helps extend the life of the battery in a system that charges and discharges frequently, but may actually slow the charging process for a small solar generator that is used infrequently.

Most charge controllers also include a **maximum power point tracking (MPPT)** feature that adjusts the voltage coming from the panel to maximize its power output. This is a nice feature, but not critical for a solar generator application.

Selecting a Charge Controller:

The process of selecting your charge controller is pretty straight forward. First, make sure that the unit is rated for the same nominal voltage as your system (in the case of a solar generator - that will very likely be 12 volts).

Then you will need to make sure it is rated to handle the maximum amps that can be generated by your solar panel.

Figure 8-2: Label from a 12 W nominal 12-volt solar panel.

The way to determine how many amps the panel may produce is to refer to the label (similar to the one illustrated in Figure 8-2) on the back of the panel.

Under standard test conditions, the Isc (Amps Short Circuit) is the most amps the panel is capable of producing under standard test conditions. But remember, solar panels can work better when it is cold, and standard test conditions measure output at 25° C (78° F).

When it is colder, the panel might generate more current. So best to give yourself a bit of room. If you multiply the Isc of the panel by 1.25, you should have more than enough "fudge factor" to deal with even the coldest day.

In the example above, the Isc for this 12-watt panel is only 0.77 amps. So even when adjusting for cold weather with a 1.25 correction factor (0.77 x 1.25 = .96) the panel will still produce less than one amp of power.

Chapter 8: The Charge Controller

Selecting a charge controller that will handle at least one amp will be adequate for a small solar generator. Larger solar panels will generate more amps, so a larger charge controller will be required.

Do not confuse the amp-hour rating from the battery bank with the amp rating required for the charge controller. The maximum amps coming from the solar panels are what determine the size (as rated in amps) of the charge controller required.

Charge Controller Connections:

Regardless of the size of the charge controller, there will likely be only three connection possibilities (and in some cases only two), as shown in Figure 8-3.

You will connect the positive and negative wires from the solar panel to the indicated connectors on the charge controller.

Leads from the charge controller to the battery are also connected as indicated.

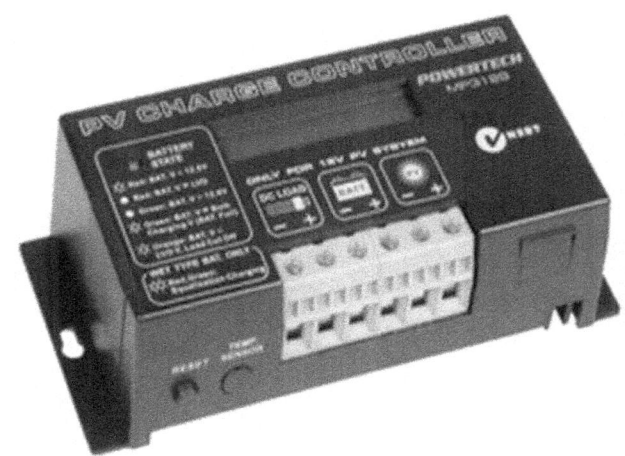

Figure 8-2:

Typical small charge controller. The unit is connected to the solar panel as well as the battery. There are also connections for optional DC loads.

There are often (but not always) connections where DC loads can be powered (from the battery, thereby avoiding the loss of power encounted within the inverter). DC loads may include a small fan (to vent the generator) or a state-of-charge meter, or even a port to charge mobile phones.

If the charge controller does not have an option to connect DC loads, you can make these connections directly to the battery.

Chapter Nine
Wiring the Generator

Note: *Care must be taken to ensure the correct size wire is used for each circuit. Overcurrent protection (fuses) will also be selected and incorporated to protect the wiring and the components from damage.*

The wiring is essentially the nervous system of the generator, moving electricity from one component to another. When items are connected together, so that electricity can flow along the **conductors**, this is referred to as a **circuit**.

The solar generator incorporates several circuits, each with their own unique wiring demands. Don't be alarmed, it is not that complicated, but it is quite important to get it right.

Wire Sizes:

In the United States, wire is measured by its gauge. It is a bit counter intuitive, but the larger the number, the smaller the wire. So #18 **AWG (American Wire Gauge)** wire is smaller than #16 AWG which is smaller than #14 AWG and so on.

Wires typically used in a solar generator will range anywhere from #24 AWG (quite small) to #0000 AWG (also referred to as 4/0 AWG) which is quite large (about the diameter of your thumb).

The larger the diameter of the wire, the more current it can safely carry. If you allow too large an electrical current to flow through too small a wire, the wire will overheat, perhaps catch on fire, break or cause some other unwanted and perhaps dangerous condition. So make sure you use the right size wire within each circuit.

You cannot get into trouble using wire that is larger than necessary (although it will cost more and perhaps be difficult to install), but you will certainly get in trouble if you use wire that is too small.

The National Electrical Code has helpfully rated various wire sizes, relating the amount of amps they can safely carry. It is the amps that impact wire size selection, the volts running through the wire have no effect (you can safely transport thousands of volts over a very small wire if the amps are low as well).

For a complete listing of wire sizes and the amps they will support, refer to the NEC Table 310.16 (if you do not have a copy of the NEC handy, this table is reproduced online in a variety of locations). In the "real world" the selection

Chapter 9: Wiring the Generator

of wire can become quite complicated. The insulation of the wire is rated for various factors such as moisture, sunlight, toxic smoke, high heat, etc.

You can also select wire made of copper or made of aluminium. Each variation will effect how much current the wire can safely transport.

In making a solar generator, we will limit the options, making it much easier to select the wire. Since we are using a small amount of wire (as opposed to hundreds of feet), we will use only copper wire. We will also assume the unit will be kept indoors, will be kept dry, and shaded (not in direct sunlight).

We will also assume there is no need to use any wire smaller than #12 AWG. Remember, you can always use wire larger than required, but not smaller. Some of the components you purchase may come pre-wired with a

Table 9-1: Ampacity Ratings of Low Heat Copper Wires
(For complete ampacity ratings, refer to NEC Table 310.16)

Wire Gauge	Maximum Current
12 AWG	25 amps
10 AWG	30 amps
8 AWG	40 amps
6 AWG	55 amps
4 AWG	70 amps
2 AWG	95 amps
0 AWG	125 amps
2/0 AWG	145 amps
4/0 AWG	195 amps

smaller wire, which is fine, as the size has been determined by engineers who understand the limits of the component and have slected a wire to match.

With these conditions in mind, refer to Table 9-1 as a guide when selecting wire sizes (we will talk about how you determine the amps of each circuit in a moment).

This table represents the lowest ampacity ratings available when using copper wires. Higher rated wires are available, but using this table gives us a "worst case" rating (an added margin of safety).

So, for example, if you have done your circuit calculations and found that the maximum current on that circuit might reach 40 amps, referring to Table 9-1 you would need to select an #8 AWG wire to ensure the electricity will be transported safely (a #10 AWG wire will only safely transport 30 amps).

Overcurrent Protection:

Even in a properly designed system, problems happen. This is the nature of things.

Sometimes too much electricity tries to flow through the circuit. In the case of a solar generator, this might happen if there is damage to the unit, or more likely, if a load that is too great is plugged into a unit not designed to handle that much power demand.

The generator's first line of defense in a case where an oversized load is connected will be the fuse or circuit breaker housed within the inverter. This comes pre-installed by the inverter's manufacturer.

CHAPTER 9: WIRING THE GENERATOR

But, if for some reason the overcurrent protection (fuse and/or breaker) in the inverter fails to work properly, we also need to install protection in the other circuits of the unit.

We can know what size fuse to select once we have determined the maximum amps that each ciruit is designed to carry. That calculation will be used to determine the size of the wire for the circuit as well. A fuse placed within the circuit should be equal to or less than the amp rating of the wire and components used in the circuit.

For example, if it was determined that a particular circuit might carry as much as 45 amps of current. Looking at Table 9-1 we see that #8 AWG wire would be too small (rated only for 40 amps) but #6 AWG wire will do nicely (rated for 55 amps). So we select #6 AWG wire for the circuit.

We could select a fuse rated at 55 amps, since the wire can handle that amount of current. However a 45-amp fuse would be the better choice, because it will protect the circuit under normal operating conditions and disconnect should something unusual happen.

You should always select a fuse or breaker at the level of amps that the circuit might carry under normal conditions. If a fuse is not commercially available at that level, select the next available size. **But never select a fuse that is rated at more than the ampacity of the wire to which it is connected.**

We could also select a lower rated fuse (say, 20 amps). This fuse will still protect the wire and equipment from over current, but will blow or disconnect at times when

Figure 9-1: PV Output Circuit

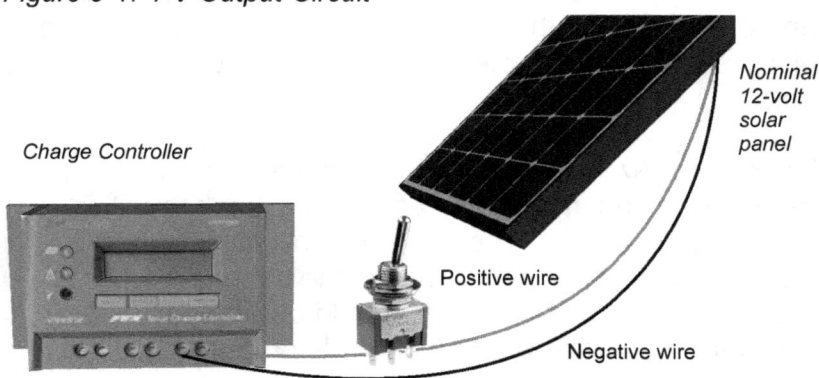

the system is operating as designed - which could prove quite aggravating.

As a rule, it is typically appropriate to select a fuse as close to the maximum designed current of the circuit, but less than or equal to the amp rating of the wire used in the circuit.

The Circuits of the Generator:

So let's now examine each circuit within a typical solar generator. These include:

- **PV Output Circuit** (from the solar panel to the charge controller).
- **Battery Input Circuit** (from the charge controller to the battery).
- **Inverter Input Circuit** (from the battery to the inverter).
- **Inverter Output Circuit** (from the inverter to the load).

CHAPTER 9: WIRING THE GENERATOR

There may be other minor circuits in the system (for meters and the like), but these are the main circuits to worry about.

PV Output Circuit:

The PV Output Circuit consists of the wires connecting the solar panel(s) to the charge controller. There is a positive wire (normally red in color) and a negative wire (normally black). Connect the positive terminal of the solar panel to the positive connector on the charge controller, and connect the negative wire in a similar manner (as shown in Figue 9-1).

The amps that will flow on this circuit are determined by the maximum amps that the solar panel can produce. The maximum amps the panel can produce under standard test conditions is the Short Circuit Amps (Isc) that will be noted on the back of the panel or in its specification sheet. We will add a safety margin by multiplying this number by 1.25.

Unless you have connected your generator to a very large solar array, the amps (even if you went to the trouble of calculating the variation due to lower temperatures), will be quite small. A typical 100-watt, 12 nominal volt panel will generate fewer than 10 amps. A smaller panel will generate even fewer amps.

So for solar panels (or arrays) smaller than 200 watts, #12 AWG wire (rated at 25 amps) should be more than adequate. The charge controller is likely fused, but if you wish to add overcurrent protection to this circuit, a 25-amp fuse will do the trick (assuming you are using #12 AWG wire).

Figure 9-2: Battery Input Circuit

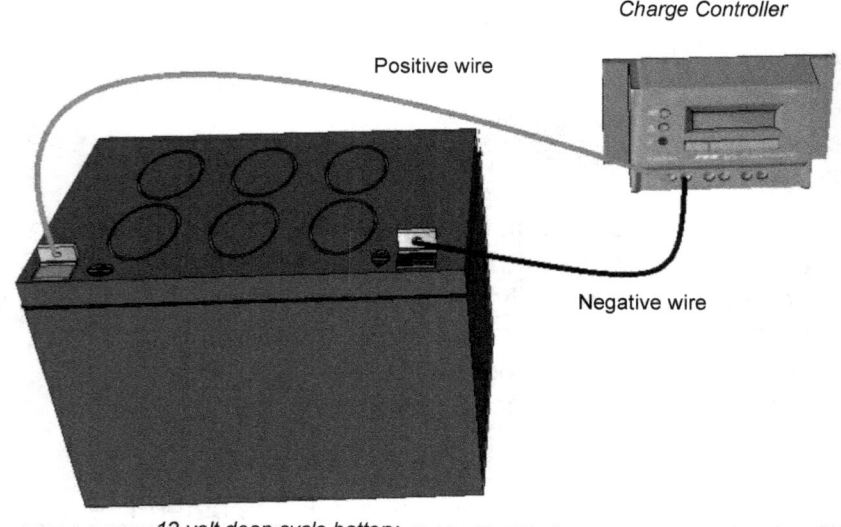

It is a good idea to incorporate a disconnect switch as close as possible to every power source. In this circuit, the disconnect should be rated as 12-volt, and rated for at least 25 amps (a disconect rated for more amps than the circuit it is connected to is fine as well, but not fewer amps).

Always place the disconnect, as well as any overcurrent protection, on the red (positive) wire. The black (negative) wire should run uninterrupted from the solar panel to the charge controller.

Summary - PV Output Circuit: For systems with a smaller than 200-watt solar panel, use #12 AWG wire, and a 12-volt, 25-amp disconnect switch. Incorporate a 25-amp fuse in the positive lead. Also make sure the charge controller is rated to handle the maximum output from the solar panel.

Battery Input Circuit:

From a very practical perspective, the Battery Input Circuit is simply a continuation of the PV Output Circuit.

This circuit connects the charge controller to the battery as shown in Figure 9-2.

Whatever was determined are the maximum amps flowing from the PV panel to the charge controller, the same calculation applies to the battery input circuit (the charge controller simply routes that energy on to the battery whenever it is needed).

If it was determined that 25 amps are sufficient for the PV output circuit, the same will apply to the battery input circuit. A #12 AWG wire will work as well for this circuit. No fuse or disconnect is required, since these were placed in the PV output circuit.

Summary - Battery Input Circuit: Continue with the same wire size as used in the PV Output Circuit. No overcurrent protection or disconnect is required in this circuit.

Inverter Input Circuit:
This circuit connects the battery to the inverter. The amps that will flow through this circuit will be determined by the load demand.

For example, if you plug in a 60-watt light bulb, the amps flowing through will be the watts divided by the volts of the circuit. Since the battery is operating at 12-volts...

amps = 60 watts / 12 volts = 5 amps

Figure 9-3: Inverter Input Circuit

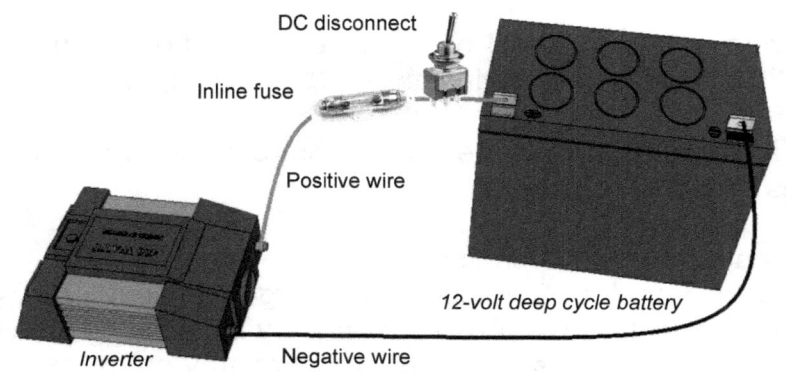

If a 1,200-watt space heater is plugged into the generator, then...

amps = 1,200 watts / 12 volts = 100 amps

Since you will likely not know the load demands that might be plugged into the generator, it is best to base the load calculations of this circuit on the maximum watts the selected inverter can handle.

An inverter with a continuous rating of 400-watts can typically manage a surge current of twice that amount (in this case, 800 watts). So ideally, the wire and overcurrent protection on this circuit should be sized to handle the surge current.

Using the example of a 400-watt inverter with a surge capacity of 800 watts, the amps that might flow over this circuit may be calculated as follows:

amps = 800 watts (inverter surge capacity) / 12 volts (nominal voltage of the battery) = 66.66 amps

CHAPTER 9: WIRING THE GENERATOR

Referring once again to Table 9-1, we find that we must use #4 AWG wire (rated at 85 amps, the #6 AWG rated at 65 amps doesn't quite do the job).

The inline fuse (overcurrent protection) incorporated into this circuit should be rated at 66 amps or less. Even though the wire can hande more than 66 amps (#4 AWG wire is rated to 85 amps), the inverter cannot. Size the fuse so it will protect the inverter.

You could get by with a smaller fuse (after all, the continuous rating of the inverter is only 400 watts, which would only require a 35-amp fuse to function). However a smaller fuse may blow under normal surge conditions.

It is important to remember that the power flowing from the battery to the inverter is 12-volts DC. The power flowing from the inverter to the load is converted to 120-volts AC. The watt rating for the loads are calculated for the higher, AC voltage. As a result, the impact (in amps) will be ten times as much for the inverter input circuit then it will be for the inverter output circuit.

Therefore, the wire size and the overcurrent protection on the inverter input circuit will be much larger than that used in the inverter output circuit.

Summary - Inverter Input Circuit: Divide the surge capacity rating of the inverter by 12 volts (the nominal voltage of the battery bank). Size the wire and overcurrent protection accordingly.

Note: For larger generators, it may be impractical to incorporate wire and fuse sizes large enough to handle the entire surge capacity of the inverter. In these systems, you can limit the current with your fuse selection. If you do this, however, you will not be able to use the full capacity of the inverter.

For example, assume you have installed a 2,500-watt inverter with a 5,000 watt surge capacity. The amp rating for the wire should be...

amps = 5,000 watts / 12 volts = 416.66 amps

Such a current would require very expensive welding cable and a very expensive fuse. It is also very unlikely that you will have an ac load that draws nearly 42 amps (5,000 watts / 120 volts = 41.6 amps) of current serviced by your generator.

Normal home wiring schemes limit circuits to 20 amps. So a solar generator capable of 2,400 watts surge capacity (watts = 120 volts x 20 amps) should be sufficient to handle most household situations.

Figure 9-4: Inverter Output Circuit

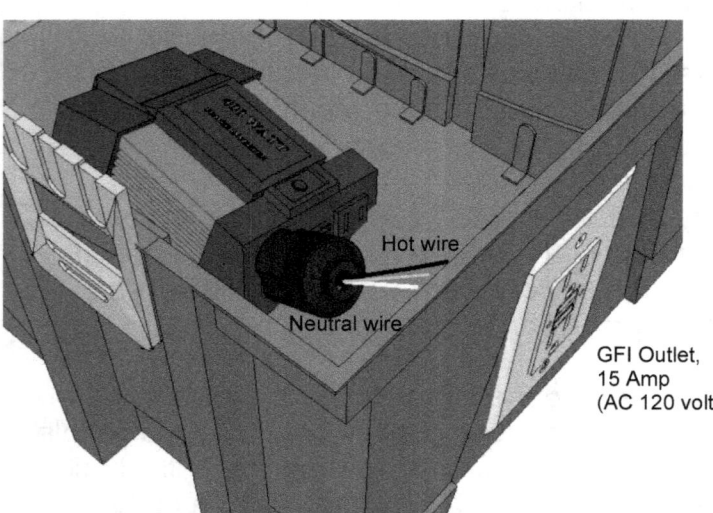

Inverter Output Circuit:

This is the circuit that flows out of the inverter to the load. In many cases people simply connect the load directly to the inverter (using the outlet supplied on the unit). If this is the case, no additional wiring is required.

When creating a solar generator, however, it is convenient to attach an electric outlet to the outside of the case, so it becomes unnecessary to leave the unit open when in use (as illustrated in Figure 9-4).

Just as with the Inverter Input Circuit, it will be the load demand that determines the current that flows through this circuit.

If it had been determined (subject to the limitations of the inverter) that this generator had a maximum capacity of 2,000 watts, then amps for the Inverter Output Circuit can be determined as follows:
amps = 2,000 watts / 120 volts (the output from the inverter) = 16.66 amps.

You will find that in sizing a 12-volt generator with 120-volt output, the amps for the inverter output circuit will always be 1/10th that of the inverter input circuit (since 120-volt output is 10 times that of 12-volt input and the watts from the load remain the same).

In this example, once again #12 AWG wire (rated to handle 25 amps) will be more than enough to handle the job.

Note that the wires in the inverter output circuit are no longer referred to as positive and negative (as was the case with the DC current that entered the inverter). AC

Figure 9-5: Various Styles of Crimp Connectors

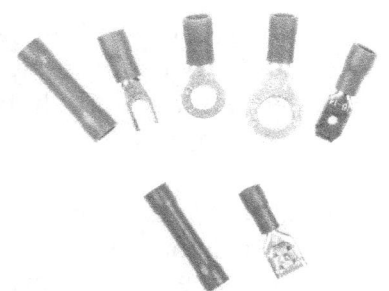

(Image from SparkFun Electronics)

circuits contain conductors referred to as the hot line (usually black) and the neutral (usually white). There will also likely be a bare copper ground wire incorporated in this circuit as well.

Brief Note on Grounding:
Normally all circuits and all metal components in an electrical system are bonded together and grounded to a metal rod driven into the earth. This allows a safe path for any stray current to find its way to earth.

Your solar generator will likely be contained in a plastic box (do not use a metal container without bonding and grounding the entire system). The mobile nature of the unit also makes grounding impractical. So bear this in

Figure 9-6:

Larger wires are inserted into the hole and clamped in place using the set screw in the top of the lug connector.

(Image from http://www.galesburgelectric.com)

CHAPTER 9: WIRING THE GENERATOR

mind - THE SYSTEM AS DESCRIBED IN THIS BOOK IS UNGROUNDED. This can present a safety issue. Refer to the NEC for proper grounding techniques.

Summary - Inverter Output Circuit: Divide the surge capacity rating of the inverter by 120 volts (the nominal voltage of the AC current leaving the inverter). Size the wire accordingly. If incorporating a GFI (recommended) outlet, make sure the amp rating on the breaker contained on the outlet has the capacity to support the amp rating of the circuit.

Connecting Wires Together:
It is impossible to anticipate all the ways the wires and components may be connected together. Smaller wires may be looped around screws and tightened into place, or attached using **crimp connectors** as illustrated in Figure 9-5.

Larger wire may require the use of **lug connectors** such as the example illustrated in Figure 9-6.

Chapter Ten
Safety

Before we pick up a screw driver, or try to connect our first wire, let's talk a bit about safety.

Remember, even low amounts of electricity can kill. According to the Center for Disease Control (CDC), exposure to as little as 1/10th of an amp for 2 seconds can cause death. Be very cautious when working with any form of electricity.

Batteries also present some unique safety concerns in addition to electric shock. So what are some safety issues associated with solar generators that we can anticipate and hopefully avoid?

Electrical Safety:
- Do not use damaged wires. Make sure all insulation is intact and clean.
- Use well insulated tools when working on the unit.
- Remove all metal jewelry or piercings that might come in contact with the unit.

Assembling a Solar Generator

- Wear ANSI approved safety glasses when working on the unit. Electricity can arc, causing eye damage and burns.
- Assume any bit of metal may be energized with electricty (whether it was intended to or not).
- Pushing too many amps over wire or components not rated to handle that current may result in excessive heat and/or fire. Always make sure the wire, overcurrent protection and components have been sized properly.
- Keep the unit dry. Water can conduct electricity and cause damage to the unit and/or injury to anyone coming in contact with it.
- People who wear pacemakers should not work with electricity, as the electromagnetic fields may interfer with the operation of the pacemaker.

Battery Safety:
- Deep cyce lead acid batteries contain acid. If spilled, they can burn your skin and/or cause serious damage to your eyes. If exposed to the acid, flush the affected skin and eyes with cool water. You can also neutralize the acid with baking soda.
- Always wear splash-resistant ANSI-approved safety goggles and electrically insulated gloves while working near batteries.
- Follow all the manufacturer's safety instructions.
- Only connect similar batteries together. Do not mix old and new batteries, wet with gel cells, or batteries of different amp-hour capacities
- Batteries are very heavy. Use proper lifting techniques to avoid injury when moving them or the solar generator unit.

Chapter 10: Safety

- Overcharged batteries can discharge hydrogen gas. This gas is explosive. Keep the unit away from flames and only operate it in a well ventilated area. Also, ensure the unit itself is well vented so that gas does not build up inside the case.
- Keep the batteries cool. It is best to store the unit in a cool, shaded place.
- Do not store the battery where it may freeze. Frozen batteries may crack, leaking acid.
- Never short the battery terminals (directly connecting the positive and the negative) with no load between these connections.
- Keep the terminals clean.

Inverter Safety:
- Inspect connections to make sure they are tight.
- Check that the inverter, as well as the load device, do not get hot during use. If the device gets too hot, disconnect it and do not attempt to use it again with the solar generator.
- Do not attempt to power any load that exceeds the capacity of the inverter. This may cause damage to the unit.
- Do not connect the unit directly to the building's electrical system. It may not meet the requirements of the building's electrical system, and may also feed power onto the grid when the grid is "down" - causing a potential for injury to those working on the lines.
- Make sure to connect the positive terminal of the battery to the positive connection on the inverter (and the negative to the negative). Do not reverse the **polarity** (negative to positive), since this will likely damage the inverter.

 ASSEMBLING A SOLAR GENERATOR

CHAPTER ELEVEN
ASSEMBLING A SOLAR GENERATOR

At this point in the process, we have learned about all the various components and how they interact within a solar generator. Now, let's put the whole thing together.

Step 1: Assess the Load Demand

Whether you are building a small solar generator or designing a complete PV system for a home or business, the first step will always be to assess the load demand of the system.

In the example we will use throughout this chapter, we will assume that the generator is designed to run a laptop computer (40 watts), an Energy Star printer (10 watts), a cordless telephone (3 watts), a lamp with a compact flourescent bulb (14 watts), and a modem (2 watts).

The purpose of the generator in this example is to continue to run a home-based business and stay connected during frequent and increasingly long power outages.

If we add all the load demands that might be drawn at any one time, we find that the grand total is 69 watts. In an effort to be more than generous with our available power, we will round up our instantaneous load demand estimate to 100 watts.

Step 2: Select the Inverter

None of the anticipated loads incorporate a motor - so surge demand is likely not a problem. They also do not require a pure sine wave signal (these loads are not electrically sensitive), so we can get by with a less expensive modified sine wave inverter.

Any modified sine wave inverter with a capacity of 100 watts or more will likely do the trick. We have searched online and found a 200-watt inverter and a 400-watt inverter. The 400-watt inverter costs only $5 more than the smaller unit, so we have decided to use the larger 400-watt modified sine wave inverter.

We might have opted to select an even larger inverter, depending on the price. You will find that the fans on smaller inverters tend to run constantly, even when powering very small loads. This can be wasteful (of energy) and a bit annoying. Larger inverters require cooling less often when running small loads. Inverters are generally more efficient if they are servicing 80% or less of their load capacity.

Chapter 11: Assembling a Solar Generator

Step 3: *Determine Length of Time the Generator can be Used before Recharging (Battery Size)*

In this scenerio, we have been experiencing frequent power outages that seem to be getting longer and longer. In the last year we have had three that lasted eight hours or more.

To be safe, we decide that we want the generator to power our little makeshift office for up to 16 hours of continuous use (without a recharge).

We know our load demand is 69 watts, but have rounded this estimate up to 100 watts.

So 100 watts x 16 hours results in a need for 1,600 watt-hours of power.

Checking our inverter, we find that it is 85% efficient. In other words, only 85% of the power it draws from the battery is converted into usable electricity. We will need to adjust the battery's capacity to account for this inefficiency. The result is:

1,600 Wh / .85 = 1,882 Wh

So because of the inefficency of the inverter, we will need a generator with 1,882 Wh capacity to provide us with 1,600 Wh of usuable power. But the capacity of a battery is measured in amp-hours, not watt-hours. So in order to convert to amp-hours, we will need to divide the watt-hours by the voltage of the battery. Thus:

1,882 watt-hours / 12 volts = 157 amp-hours

If we drained our battery down to nothing, we could get by with a 157 Ah rated battery. However we will likely not drain it to zero. In fact, our inverter is designed to turn off when the battery reaches a 90% depth of discharge (only 10% of the charge remaining). With this in mind, we will need to divide 157 Ah by .90 (as we can only use 90%) to find:

157 Ah / .90 = 174 Ah

We now know that we need a 174-Ah battery to power this load with this inverter for 16 hours before recharging.

So we go out and purchase a 12-volt, 175-Ah deep cycle battery.

Remember to provide a vent in the generator case so that any gasses that might build up if the battery overcharges can disapate. Also, the inverter will need a constant airflow to remain cool during operation.

Step 4: *Determine Length of Time the Generator can Recharge Between Uses (Solar Panel Size)*

Normally a solar generator will sit idle for several weeks between uses. It is an emergency power source, not designed for everyday use. As a consequence, the solar panel used to recharge the unit does not have to be very large. But there is still a bit of an art in sizing it properly.

First, we must determine how many amp-hours within the battery will need to be recharged after each use.

Chapter 11: Assembling a Solar Generator

In our example, where we are using a single 175-Ah battery, operating to a 90% depth of discharge (DOD) before the unit shuts down.

So worse case scenerio, we can assume 175 Ah x .90 = 157.5 Ah recharge each cycle. Our system is operating at 12 volts, so 157.5 Ah x 12 V = 1,890 Wh required to recharge our system.

Detailed steps on sizing the panel can be found in Chapter 7 of this book. For the purposes of this example, we shall assume the following conditions:

- We are satisfied with a two-week period of time (14 days) to recharge the generator after each use.
- At our location, we receive on average about 4 hours of peak sunlight per day.
- We are able to orient our solar panel to the south (assuming we are in the northern hemisphere).

Sizing the panel with these assumptions is a pretty straightforward process.

14 days x 4 hours of peak sunlight per day = 56 hours of peak sunlight during the designated period.

1,890 watt-hours / 56 hours = 33.75 watts

So we will need a panel rated at 33.75 watts or greater (probably a 35-watt panel).

A larger panel will reduce the number of days required to recharge the generator. For example, if we happened to get a good deal on a 255-watt panel, then...

1,890 watt-hours / 255 watts = 7.41 hours / 4 peak hours of sunlight per day = 1.85 days

So with the much larger panel, we can reduce the amount of time require to recharge the battery after each use from 14 days to less than two days. Just make sure the larger panel is also rated at 12 nominal volts.

Step 5: Select the Charge Controller

Once you have selected your solar panel, look at its specifications to find the short circuit current (Isc) rating. This is the theoretical highest current (amps) the panel can produce under standard test conditions.

There are a number of rather complex calculations you could make to adjust this number for various worst case temperature conditions, but you will be safe if you simply multiply the Isc by 1.25 (giving it a 25% fudge factor).

So if the Isc for your panel was 4.56 amps, adjust that by multiplying by 1.25, giving: 4.56 amps x 1.25 = 5.7 amps

So any 12-volt charge controller rated to handle 5.7 amps or more will do the trick.

Step 6: Sizing the Wiring System

The next step will be to select the wire that will connect everything together. Remember that there are various circuits within the system, and not all require the same size wire.

CHAPTER 11: ASSEMBLING A SOLAR GENERATOR

The circuits will include:

- PV Output Circuit (from the solar panel to the charge controller) illustrated in Figure 11-1
- Battery Input Circuit (from the charge controller to the battery) illustrated in Figure 11-2
- Inverter Input Circuit (from the battery to the inverter) illustrated in Figure 11-3
- Inverter Output Circuit (from the inverter to the load) illustrated in Figure 11-4

PV Output and Battery Input Circuits:
The same equation used in calculating the amps flowing through the charge controller can be used to size the

Figure 11-1: PV Output Circuit

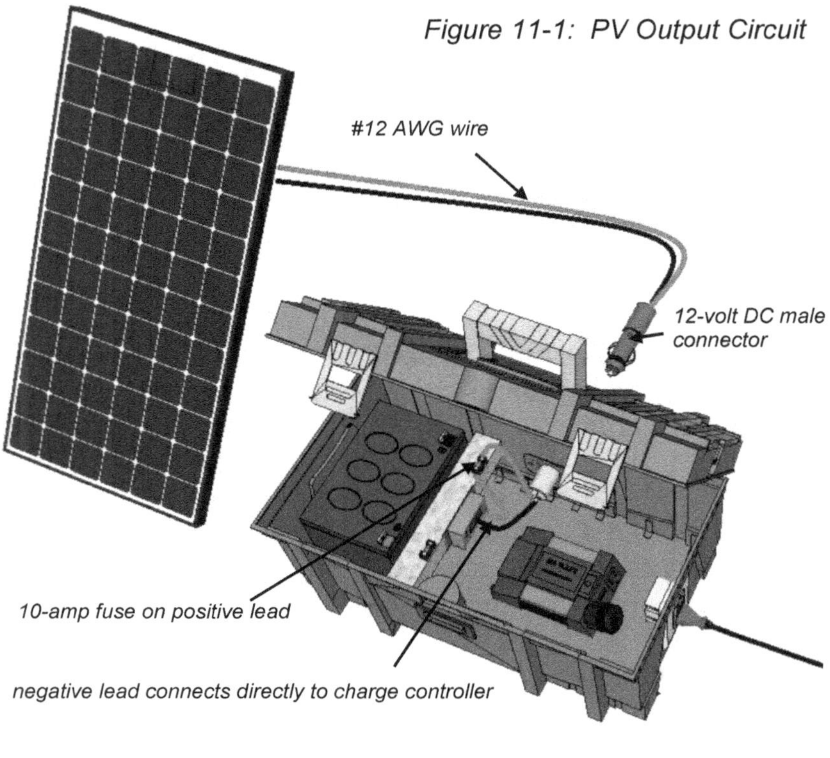

#12 AWG wire

12-volt DC male connector

10-amp fuse on positive lead

negative lead connects directly to charge controller

Figure 11-2: Battery Input Circuit

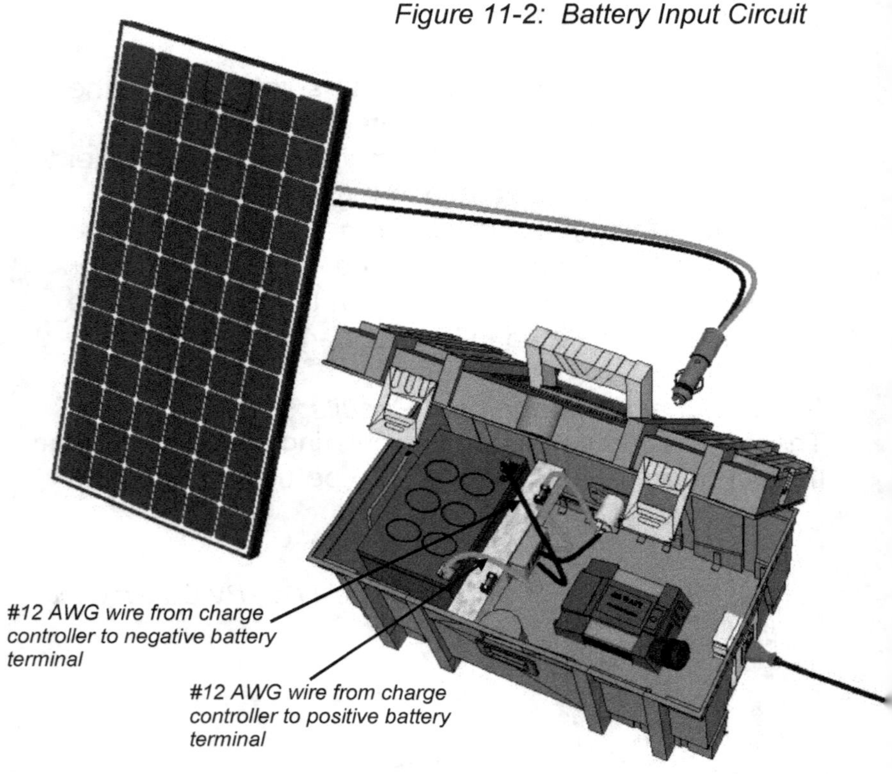

#12 AWG wire from charge controller to negative battery terminal

#12 AWG wire from charge controller to positive battery terminal

wire for both the PV Output Circuit and the Battery Input Circuit.

In our example, we have already calculated that the solar panel can generate as much as 5.7 amps (its Isc plus our 25% fudge factor).

So we will need to select wire that can hande at least 5.7 amps. This is not a lot of current, and a review of Table 9.1 shows that #12 AWG wire can handle up to 25 amps (more than enough).

We do not recommend using wire smaller than #12 AWG, since it can be hard to work with, and will save only

CHAPTER 11: ASSEMBLING A SOLAR GENERATOR

pennies on the project. So better to go with the more robust wire.

You will want to purchase black wire for the negative lead of each circuit, and red wire for the positive lead of each circuit.

It is also nice to be able to connect and disconnect the solar panel from the unit. A male and female 12-volt DC male and female "cigarette lighter" connector works well for this purpose. More on that later, but this disconnect will be incorporated into the PV Output Circuit.

You may also want to include a fuse in the positive lead from the PV panel to the charge controller (the PV Output Circuit) to protect the charge controller and battery in case stray current somehow finds its way onto the wire (lightning, a short circuit, static charge, etc). The fuse should be sized just above the expected current (in this case 5.7 amps) but must be equal to or less than the capacity of the wire (in this case 25 amps). A 7-amp fuse in this example would be ideal, but a fuse up to 25 amps would work as well.

Inverter Input Circuit:
The wire size for the Inverter Input Circuit is determined by the load demand that might be attached to the generator. In this example, we have decided that the load demand we expect to service is only 100 watts (determined in step one).

However we have selected a 400-watt inverter with a surge rating of 800 watts. So even though we think we will only use 69 watts at any given time, the inverter is capable of generating 800 watts (even though it is only for a few seconds).

Figure 11-3: Inverter Input Circuit

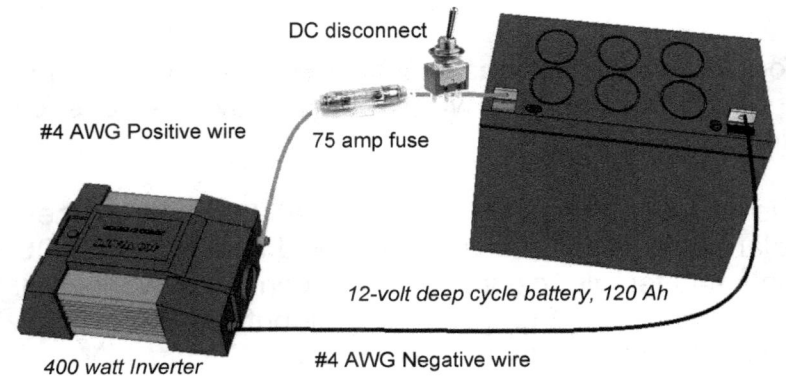

So we should size the wire for the Inverter Input Circuit and the Inverter Output Circuit based on the "worse case" possiblity of 800 watts (the inverter will malfunction if we try to generate more watts than that).

The Inverter Input Circuit operates at 12 volts (coming from the battery to the inverter). As a result, the circuit may be asked to carry up to 800 watts / 12 volts = 66.67 amps of current.

Looking at Table 9.1, we see that we will need to use at least #4 AWG cable (rated to 70 amps) to safely transport that amount of current.

We will also need to incorporate a fuse into the positive lead of this circuit to protect the inverter, should something go wrong. In this case, the fuse should be rated at 70 amps (the next available standard fuse size over 66.67 amps, the maximum current the circuit is expected to carry under normal operating conditions).

A smaller wire and fuse can be used, but this will limit the functional capacity of the inverter.

CHAPTER 11: ASSEMBLING A SOLAR GENERATOR

Inverter Output Circuit:
From the inverter to the load, however, the voltage of the circuit has been increased from 12 volts to 120 volts. It is also AC current, rather than DC current (the purpose of the inverter, after all).

Most inverters give you the option of plugging the load directly into the inverter. However, you will likely find it convenient to place a regular 120-volt outlet on the exterior of the generator, so that you can easily plug loads in and remove them, without opening the generator case.

If you extend the inverter output circuit (as shown in Figure 11-4), the maximum amps flowing over this circuit will be:

800 watts / 120 volts = 6.67 amps

Using #12 AWG wire will once again be more than enough for this circuit (rated to 25 amps).

Figure 11-4: Inverter Output Circuit

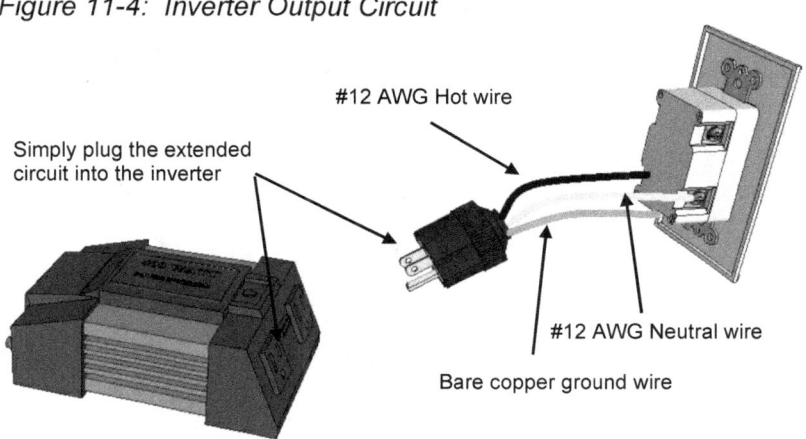

This is now an AC circuit, so it is a good idea to follow the color coding for AC. With this in mind, the "hot wire" will be black, the "neutral" will be white. Also incorporate a bare copper wire for the ground.

Also note that the black wire generally connects to the bronze screw (in the outlet or switch) and the white wire connects to the silver screw. This avoids reversing them and thus reversing the polarity of the circuit.

If incorporating a GFI outlet, the breaker in the outlet will serve as overcurrent protection for this circuit.

Optional State of Charge Circuit:
After using your generator for a period of time, you will find that it is quite helpful to, at a glance, be able to determine the state of charge of the battery. Inexpensive digital meters are available that can be incorporated into the generator that will accomplish this purpose.

Figure 11-5:

Blocking Diode in junction box

Chapter 11: Assembling a Solar Generator

If your charge controller has the option to connect DC loads, this is an ideal place to hook up the meter. Since the battery powers any DC loads, the meter will measure the state of charge of the battery (rather than the input from the PV panel).

If your charge controller does not have the DC load option, you can connect the meter directly to the battery terminals. You will also want to incorporate a DC disconnect switch in this circuit so that you can turn off the meter when it is not needed. Otherwise the meter will slowly, but constantly, drain just a bit of energy from your battery.

Optional DC Disconnect in PV Input Circuit:
It is always a good idea to incorporate a disconnect switch as close to any power source as possible. While the PV panel will only generate a small amount of energy, it is still a power source.

A simple disconnect solution for this circuit is to simply unplug the panel from the generator. Crude, but effective.

Also, many male DC "cigarette lighter" connectors come with a built in "on-off" switch. This can serve as the DC disconnect switch for this circuit. Note, however, that many of these switches also incorporate an LED light to indicate if they are "on" or "off."

However, since the male adaptor is connected directly to the PV panel (a power source), the indicator light will always be lit when the panel is exposed to light. So do not rely on the indicator light as an indication as to whether or not the switch is in the closed position.

One reason a disconnect is helpful in the PV Output Circuit is that very inexpensive solar panels occassionally will not incorporate **blocking diodes**. These diodes (located in the junction box of the solar panel as illustrated in Figure 11-5) prevent power from draining from the battery through the panel.

At night a solar panel can act like a load, draining a small amount of power from the battery. A blocking diode ensures that the flow of power is only one way - from the panel to the battery.

If your panel does not incorporate a blocking diode, it may be helpful to disconnect or turn off the circuit when the panel is not exposed to light.

Chapter Twelve
A Picture Is Worth a Thousand Words

Now that we have studied all the theory behind the construction and operation of solar generators, let's put one together. We will start small, with a 400-watt unit that is designed to operate small loads for only a short period of time.

You may, of course, wish to design and build a much larger unit. The processes, whether a small unit like this example or a huge unit designed to power a cellular tower (one example of how these are employed), are exactly the same. Just bigger components and larger wires.

For this example, we will provide details as to where they were purchased and how much was paid for each of the components (as of January 2015). In this way you can get a "real world" sense of costs and perhaps even build this exact unit.

Assembling a Solar Generator

Components may be available at multiple sources, we are simply providing a list of where we purchased them (not an endorsement of quality or anything like that).

Table 12-1: 400-Watt Generator Components

#	Item	Where	Cost
1	400 Watt Inverter	Harbor Freight	$19.95
1	7-amp 12-volt Charge Controller	Home Depot	$17.00
1	22 inch tool box	Home Depot	$14.97
1	30 Watt, 12-volt Solar Panel	Amazon.com	$57.99
1	State-of-charge meter	Amazon.com	$6.45
1	DC 12-volt rocker switch	Amazon.com	$2.45
1	35 Amp-hour, 12-volt Battery	Harbor Freight	$69.99
1	Male Car Power Adapter	Radio Shack	$7.19
1	Female Accessory Outlet	Radio Shack	$7.19
2	Soffit Vents	Carter Lumber	$2.50
1	40-amp fuse & holder	Amazon.com	$3.05
1	GFI Duplex Outlet	Home Depot	$11.97
1	Duplex Outlet Box	Home Depot	$0.49
1	Male Electric Plug	Home Depot	$3.48
1	Outlet Cover	Home Depot	$0.49
4 ft	10 AWG wire	Home Depot	$0.80
	Wire Connectors	Home Depot	$2.00
	Total Cost:		**$ 227.96**

Chapter 12: A Picture is Worth 1,000 Words

Some of the items used to build the generator, such as wire, connectors, even bits of wood and screws, are assumed to be lying around in your garage somewhere. Plus this number is not inclusive of sales tax and shipping. So perhaps a better estimate of cost for this unit is somewhere between $250 - $300 (not counting your labor, of course).

Completed 400-Watt Generator

A compact little solar generator that will provide up to 400 watt-hours of power between charges. This generator has enough capacity to power a laptop, modem, cordless telephone, desk light and printer for about 5 hours. The unit will then fully recharge when exposed to 12 hours of full sunlight.*

Assembling a Solar Generator

*In this example, the laptop draws 40 watts, the printer 10 watts, the desk lamp 13 watts (compact fluorescent bulb), the modem 2 watts and the cordless phone 2 watts (for 67 watts total). The battery is drained to a 90% depth of discharge.

Step 1: Assemble the Components and Tools

<u>Tools Required:</u>
- Electric Drill
- Razor Knife
- Crimping Tool
- Wire Strippers
- Screwdrivers
- Multimeter
- Crescent wrench
- Linesman Pliers
- Needle Nose Pliers
- Drill Bits
- Marker
- 2" hole bit
- Electrical Tape

CHAPTER 12: A PICTURE IS WORTH 1,000 WORDS

Step 2: Prepare the Box

Drill two holes for the vents, one on each end of the box. This will allow for good ventilation for the inverter's fan and to avoid the build-up of gases from the battery.

Clean the hole with the razor knife

Place the vents (they should click into place, but you may wish to lock them in with silicon caulk)

ASSEMBLING A SOLAR GENERATOR

Trace the GFI outlet (they come in different sizes)

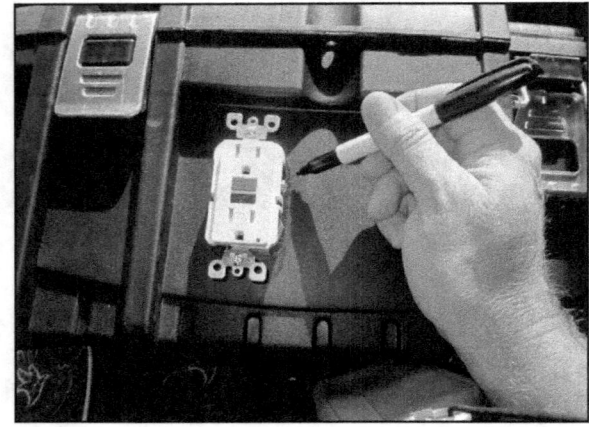

Cut out for the outlet. Don't forget to drill holes for the screws that will fasten the outlet to the plastic box (mounted inside).

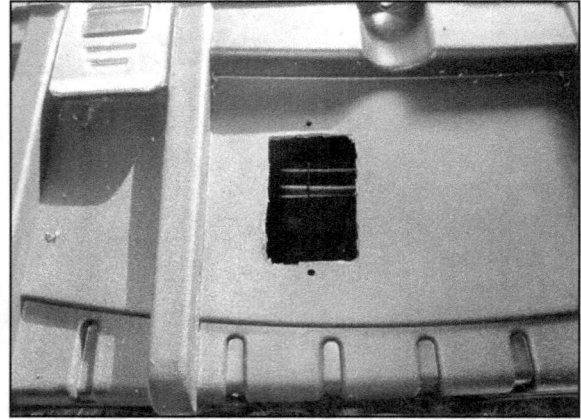

Repeat the process (place, trace, cut) holes for the volt meter, the female accessory port, and the rocker switch.

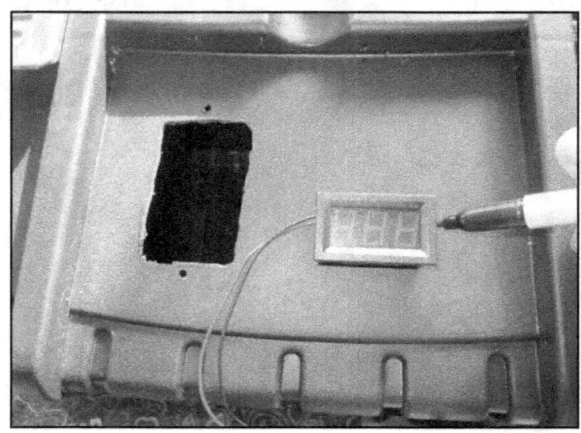

CHAPTER 12: A PICTURE IS WORTH 1,000 WORDS

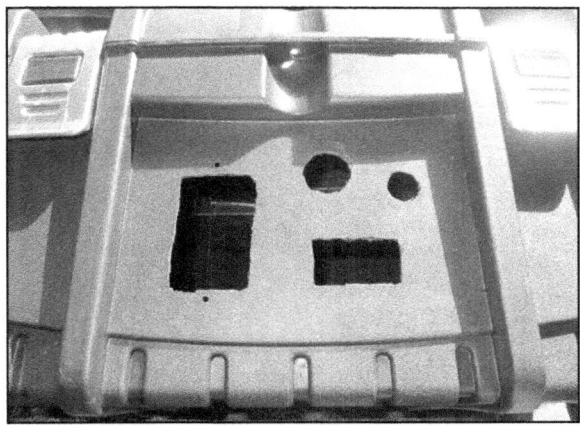

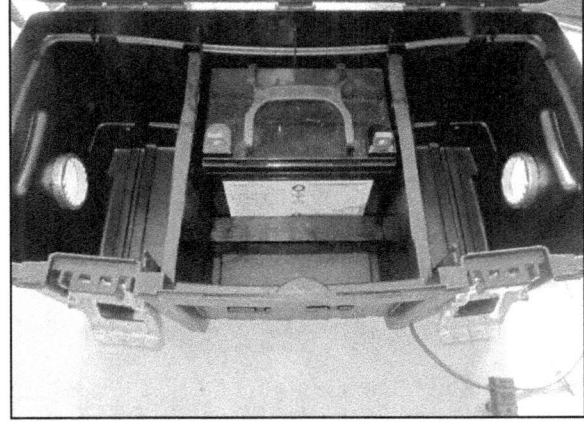

The holes are prepared and awaiting all the bits.

You will likely wish to secure the battery in some manner. Here we have placed two pieces of 3/8 inch plywood on either side, secured to a 2x2 bit of wood in the center bottom. We have found that placing the battery in the middle of the unit makes it much more balanced to carry. The plywood will also give a purchase point to mount the charge controller.

Step 3: Installing the Charge Controller

Mount the charge controller to the right hand bit of plywood. This charge controller is sized to handle the selected solar panel. Our panel is 30-watts with an Isc of 1.83 amps (so a 7-amp controller is more than robust enough).

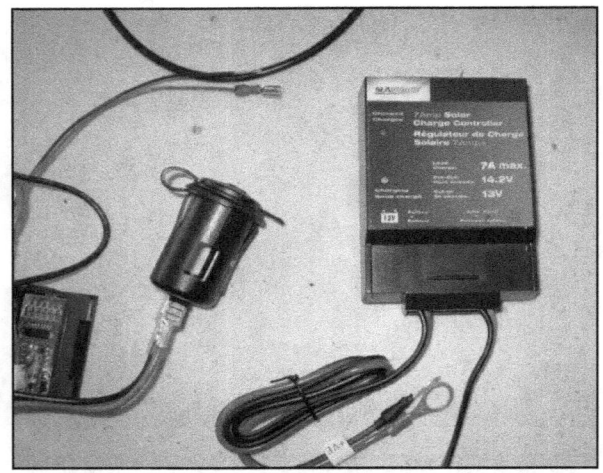

Connect the leads from the solar panel connection of the charge controller to the back of the accessory outlet. The positive lead will connect to the center tab of the outlet. The accessory outlet should simply click in place from the outside, if the hole was drilled properly.

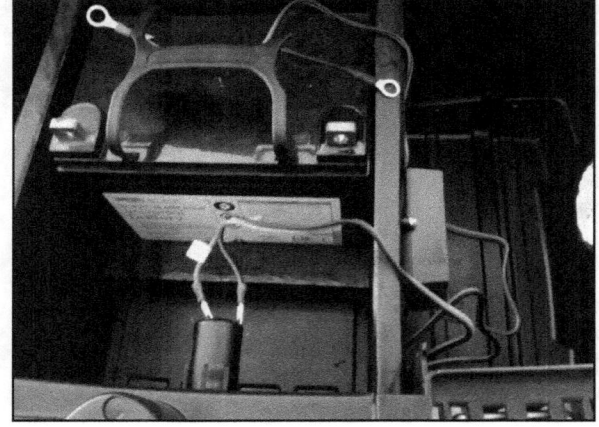

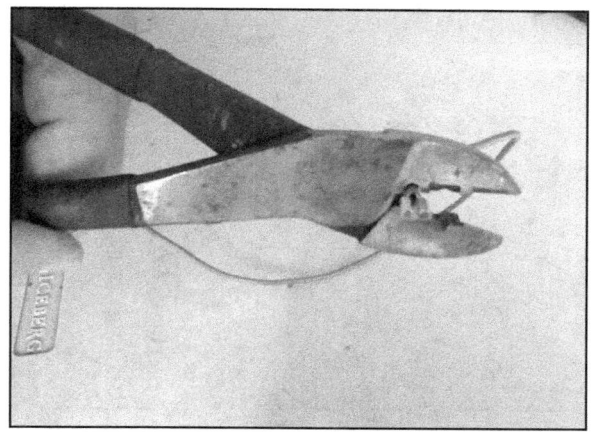

You will likely have to strip the wires and crimp on female spade connectors to attach to the outlet. Ring connectors can be used to connect the charge controller to the battery.

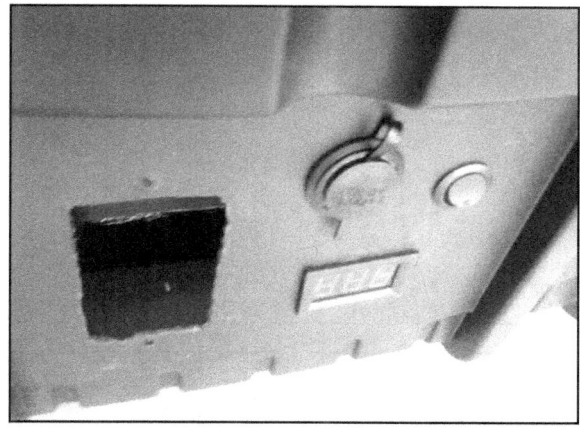

Step 4: Installing the Voltage Meter

Snap the meter and the rocker switch into place from the outside.

Connect the red lead from the meter to the switch, then from the switch to the positive terminal of the battery. The black lead from the meter connects directly to the negative terminal of the battery.

Assembling a Solar Generator

Step 5: Installing the Inverter

Most inverters come with wires already sized to handle the amp capacity of the unit. We will use this wire when connecting to the battery. It may be necessary to replace the clamp connectors with ring connectors for a secure fit.

The negative (black) lead from the battery will connect directly to the negative connection on the inverter.

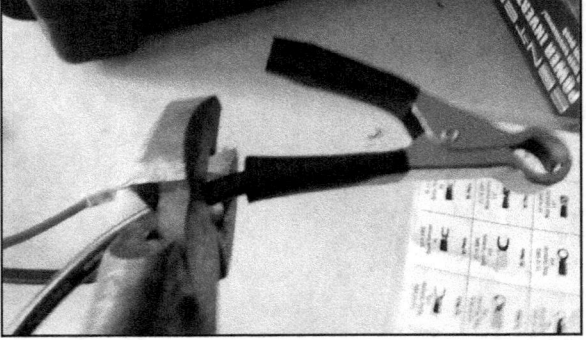
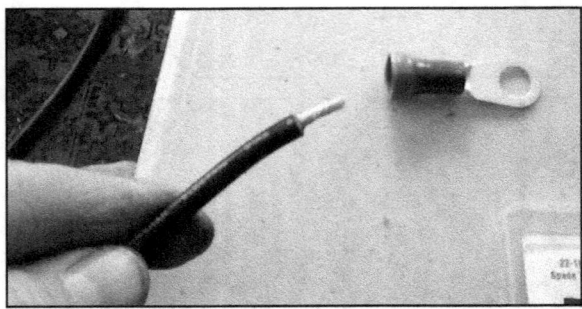

Chapter 12: A Picture is Worth 1,000 Words

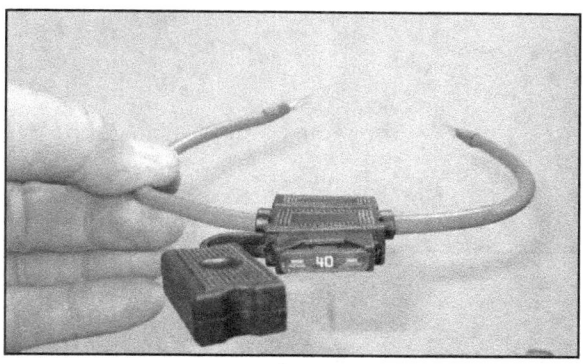

A fuse must be incorporated into the positive lead from the battery to the inverter to protect the components and wires. In this case we have used a 40-amp* standard automobile fuse.

400 watts (inverter) / 12 volts (battery) = 33.33 amps

The wire used is #10 AWG rated at 40 amps, so the fuse cannot exceed that amperage.

*If you have been paying attention, you may wonder why we selected only a 40-amp fuse. After all, the inverter has a surge capacity of 800 watts, so:

800 watts / 12 volts = 66.67 amps

But we will not be using this unit with loads requiring a surge, and therefore we will not need to use a wire larger than #10 AWG. We have, however, effectively limited our inverter's capacity to around 450 watts.

A disconnect switch might also be incorporated into the positive circuit from the battery to the inverter (not included in this example)

Step 6: Extending the Inverter Output Circuit

In order to access the power from the outside of the box, we will extend from the AC outlet built into the inverter to a GFI duplex outlet mounted on the outside of the box.

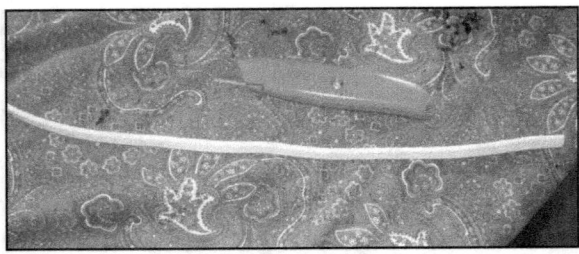

Using a bit of Romex (#12 AWG interior household wire), strip off insulation, then terminate wires into male outlet plug.

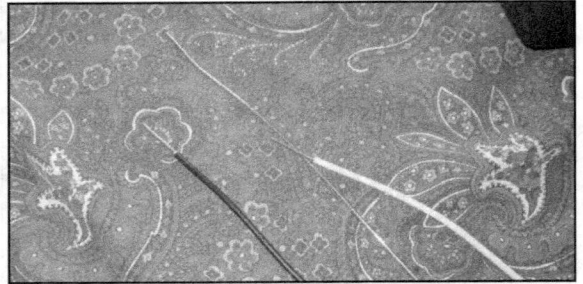

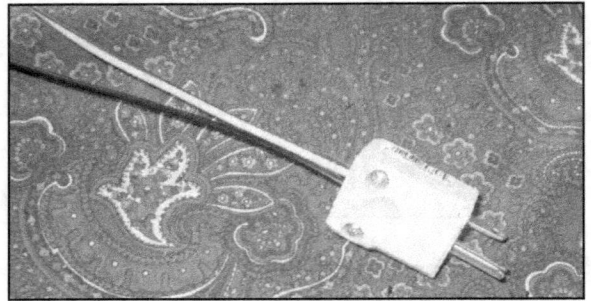

Feed the wires through the outlet box, then connect the GFI duplex outlet.

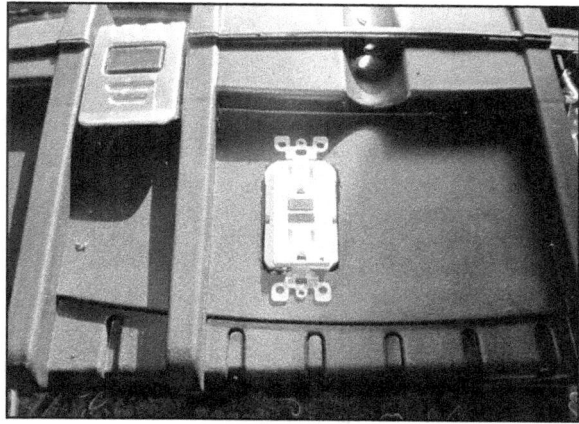

From the INSIDE of the box, feed the GFI outlet out through the hole (the plastic box and the male plug will remain inside).

Line up the plastic box and the outlet, and use the screws provided with the outlet to secure the GFI outlet to the plastic box.

Plug the male electrical plug into the inverter.

Make sure the GFI breaker (on the front of the outlet) has not tripped during shipping or installation.

Attach the duplex outlet face plate.

Assembling a Solar Generator

Step 7: Connect the Solar Panel

Your solar panel may arrive with no connectors attached, or with MC4 connectors such as the panel in this example.

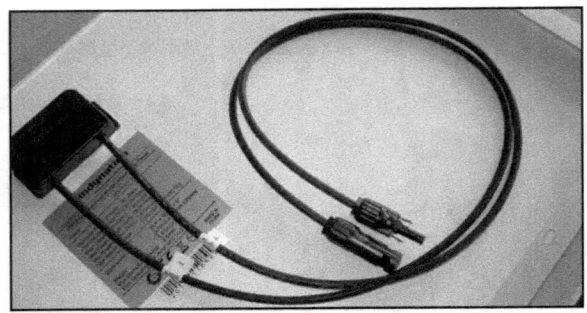

You will need to cut off the installed connectors, and attach the male 12-volt car power adapter. Note that this adapter has a built-in switch*, which is handy for disconnecting the power of the panel from the unit.

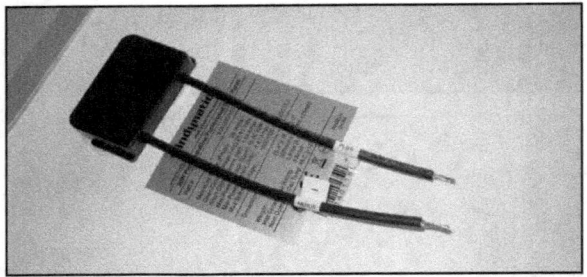

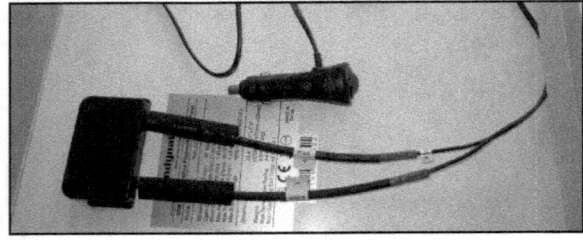

After connecting the wires (make sure the positive lead from the panel is connected to the positive lead in the adapter), weather proof with electrical tape or heat shrink tubing.

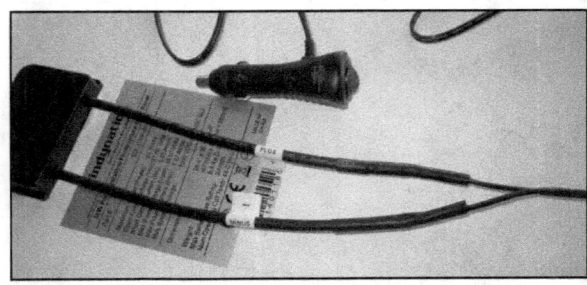

*If the switch has an LED indicator light, you will find that it will remain lit whenever the panel is exposed to light, even when in the "off" position.

Chapter Thirteen
Maintenance

Every tool must be properly maintained, and your solar generator is certainly a tool that you will want to use for many years. Assuming it is not physically damaged, most of the components in the generator should last for a decade or more.

Solar Panel Maintenance:

Solar panels are extremely robust and designed to last 25 years or more with little or no maintenance. If the panel is permanently mounted outside, a few periodic tasks should help in maintaining a quality performance from this component. These include:

- Wash dust from the panel and dry it with a soft cloth.
- Remove snow whenever necessary.
- Make sure the panel is always in full sunlight, avoid shading if at all possible.
- Periodically check to make sure all the connections are secure and the wires have not been damaged.

Battery Maintenance:

The battery is the most vulnerable part of the entire generator. In order to get many years of life from it:

- Periodically inspect the wire connections to ensure they are tight.
- Keep the terminals clean (free from any sulfur buildup).
- Check the water levels periodically (if using an unsealed lead acid battery). Fill to the level indicated with distilled water as necessary.
- Once a year or so, equalize the battery. This can be done by hooking it up to an AC-powered battery charger and fully (and perhaps over) charge the battery. This process cleans the cells within the battery.
- Store the battery in a cool, shaded area.
- Prevent the battery from freezing.

Inverter, Charge Controller Maintenance:

The electronic components within the solar generator will likely either work, or not. Do not attempt to open any of these components should they fail. Replace them or have them serviced by the manufacturer. They typically contain no serviceable parts.

Periodically check all wiring connections, inspect the units for cracks or other damage. Inspect the wiring to ensure it is not damaged as well.

Chapter Fourteen
Troubleshooting

As you assemble your solar generator, you will likely encounter some problems. While we cannot anticipate everything that might go wrong, we can try to anticipate some of the more common challenges.

Lower than expected power output from the solar panel:

The charge controller should maintain a voltage range from the solar panel close to the panel's Vmp (voltage maximum power point) when the panel is exposed to full sun. This can be measured with a multi-meter by testing the connections from the solar panel at the charge controller.

If lower than expected, the panel may be shaded. The angle of the panel is not set at 90 degrees towards the sun. The wiring connections may be loose.

In very hot weather, some panels experience heat fade (they lose capacity due to excessive heat). In this case, shade the panel until cool (the situation should then cor-

rect itself. Do not try to cool the panel with water (may cause it to crack).

Sudden system failure:

If the generator suddenly stops working, there are a number of possible causes. These include:

- A wire has become disconnected. Make sure all circuits are intact and all connections are secure.
- The batteries are dead. Normally the inverter will sound an alarm when this occurs, but not always. The state-of-charge meter should indicate that the system is operating at or near 11 volts if this is the case.
- The inverter has stopped working. If the indicator light (most units have them) is no longer green but some other color, refer to the inverter's manual to determine the cause. Typically either there is not enough power in the battery to run the inverter, the load demand is greater than the capacity of the inverter, or the inverter has gotten too hot. Disconnect all loads, turn off the inverter, and let it rest for at least 30 minutes. Make sure the cooling fan is not blocked and there is plenty of ventilation in the box. If the problem continues (after ensuring the load is properly sized to the inverter), you may need to replace the inverter.
- Make sure the inverter is switched on.
- Make sure the battery is not disconnected (if a battery disconnect has been installed).
- Check all fuses to see that they are not "blown".
- Check the GFI breaker on the outlet to ensure that a ground fault has not caused it to disconnect.

CHAPTER 14: TROUBLESHOOTING

Some GFI breakers are sensitive to movement, and may trip when the generator is relocated or bumped.

There is static on the screen or from the speakers I am powering with the generator:

If static or noise is coming from a television, computer or audio device powered from the solar generator, it is likely that that device requires a pure sine wave inverter to function properly. If you have installed a modified sine wave inverter, you will need to upgrade it for the device to operate properly while connected to the inverter.

The state-of-charge meter is giving me a false voltage reading:

First, test the state of charge with a multimeter to make sure the reading is in fact false. If so, you may have a faulty meter and it will need to be replaced. This can be tested by connecting the meter directly to the battery (positive lead to positive battery terminal, negative lead to negative battery terminal) and see if the reading now matches the reading obtained from the multimeter. If so, you may have connected to the wrong connection points on the rocker switch. Try changing the connections on the switch until a proper voltage reading is obtain on the state-of-charge meter.

Assembling a Solar Generator

The generator will not service a load as long as it used to:

Most likely the problem lies with the battery. Check the terminal connections, and check the water levels. Equalize the battery by overcharging it with an AC powered battery charger. Also, batteries will lose up to 25% of their capacity if stored in too cold an environment.

After the battery has been fully charged (either with the solar panel or a battery charger), test the voltage to ensure it is at about 12.7 - 12.9 volts. Let the battery sit for a day (with no load attached). Check to see that the voltage has not decreased. If the battery is losing voltage when not in use, it may need to be replaced.

If the battery does not achieve a full state of charge after being exposed to ample sunlight, the charge controller may have become defective and will need to be replaced.

The wires in the unit have become very hot or melted:

The wire gauge is either too small for the circuit, or the fuse is too large and has not protected the system properly. Check to make sure all wire gauges and overcurrent devices have been sized properly.

FIGURES & TABLES

FIGURES

page 12.... *Figure 4-1:* Typical Stand-Alone PV System
page 14.... *Figure 4-2:* Typical Solar Generator Components
page 18.... *Figure 5-1:* Typical Energy Use Label
page 20.... *Figure 5-2:* Kill A Watt Meter
page 22.... *Figure 5-3:* A toaster drawing 841 watts when running
page 25.... *Figure 5-4:* DC and AC electrical signal waveforms
page 26.... *Figure 5-5:* Square wave, modified sine wave and sine waves
page 31.... *Figure 6-1:* Typical "small" deep cycle battery.
page 32.... *Figure 6-2:* Typical "mid-sized" deep cycle battery.
page 38.... *Figure 6-3:* Four 12-volt, 90 amp-hour batteries connected together in series.
page 38.... *Figure 6-4:* Four 12-volt, 90 amp-hour batteries connected together in parallel.
page 44.... *Figure 7-1:* Typical monocrystaline panel
page 45.... *Figure 7-2:* Typical polycrystaline panel
page 45.... *Figure 7-3:* Typical amorphus panel
page 47.... *Figure 7-4:* Mount the solar panel outdoors
page 48.... *Figure 7-5:* How the sun's angle effects power
page 49.... *Figure 7-6:* Altitude angle throughout year
page 56.... *Figure 8-1:* Small 12-volt series charge controller
page 58.... *Figure 8-2:* Label from a 12 W 12-volt solar panel.
page 59.... *Figure 8-3:* Typical small charge controller.
page 66.... *Figure 9-1:* PV Output Circuit
page 68.... *Figure 9-2:* Battery Input Circuit
page 70.... *Figure 9-3:* Inverter Input Circuit
page 72.... *Figure 9-4:* Inverter Output Circuit
page 74.... *Figure 9-5:* Various Styles of Crimp Connectors
page 74.... *Figure 9-6:* Lug Connector
page 87.... *Figure 11-1:* PV Output Circuit
page 88....*Figure 11-2:* Battery Input Circuit
page 90....*Figure 11-3:* Inverter Input Circuit
page 91....*Figure 11-4:* Inverter Output Circuit
page 92....*Figure 11-5:* Blocking Diode in junction box

TABLES

page 21.... *Table 5-1:* Appliances that typically draw less than 400 Watts
page 23.... *Table 5-2:* Appliances that Incorporate Motors
page 24.... *Table 5-3:* Appliances that Generate Heat
page 33.....*Table 6-1:* Voltages at various Depth of Discharge
page 63.....*Table 9-1:* Ampacity Ratings of Low Heat Copper Wires
page 96.....*Table 12-1:* 400-Watt Generator Components

ASSEMBLING A SOLAR GENERATOR

OTHER BOOKS FROM BRS Press

***Understanding Photovoltaics:
An Easy-to-follow Study Guide for Solar Electric Certification Programs***

At last, here it is, available for the first time to those not enrolled in our photovoltaic training courses. Field tested by scores of students, this easy-to-follow text is designed to take an extremely "non-technical" student with zero background in PV, and literally teach them how to design and install a variety of residential systems.

This text is also designed to help prepare students who wish to sit for industry-accepted certification programs such as the ETA Level 1 PV Certification or the NABCEP PV Certificate Program. Each chapter contains multiple choice review questions, as well as labs that comply with "hands-on" requirement of the ETA certification.

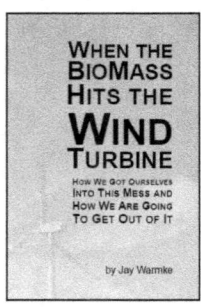

***When the BioMass Hits the Wind Turbine:
How We Got Into This Mess and How We Are Going to Get Out of It***

The world as we know it is about to end... again. And it will be great! Society is at a tipping point where the age of fossil fuels is coming to an end. Amazing changes are in store, and much faster than you might imagine.

This easy-to-read, fact-filled book walks the reader through the REAL history of energy (did you know the light bulb was already 80 years old when Thomas Edison got his hands on it?). It takes an objective look at the present energy situation, and then makes 10 stunning predictions that will change our energy future forever.

Available at www.bluerockstation.com or at www.amazon.com

Index

A
- AC load 13, 14
- alternating current (AC) 13, 25
- altitude 49-50
- American Wire Gauge (AWG) 62
- amorphus 44, 45
- amps 4, 19-20, 30, 31, 51, 57-59, 62-65, 67-73, 78, 86-91, 102, 105
- array 11-12, 14-15, 41, 43-44, 55, 67
- azimuth 49

B
- battery bank 11-15, 27, 29-31, 33-35, 37-39, 41, 44, 50, 52-53, 55, 59, 71, 75

C
- charge controller 11, 12, 14-15, 55-57, 59-60, 66-69, 86-89, 92, 93, 96, 101-103, 110-111, 114
- circuit 4, 20, 58, 61-62, 64-71, 73-75, 86-94, 105-106, 114
- compact florescent 9, 21
- conductors 61
- continuous capacity 22, 23

D
- DC load 13-14, 93
- deep cycle batteries 13, 31
- depth of discharge 33-34, 84, 85, 98
- direct current (DC) 12, 24

E
- Energy-Star 9

G
- gel battery 40-41, 78
- glass matt battery 31, 40-41

H
- hertz 18, 20, 24

I
- insolation 50, 51
- inverter 11-15, 17, 19-21, 23-27, 29-30, 32, 34, 36-37, 41, 64, 65, 67, 69-75, 79, 82-84, 87, 89-91, 96, 104-107, 110, 112-113
- irradiance 50

K
- Kill A Watt meter 20
- kilowatt-hours (kWh) 4, 30

L
- LED bulbs.. 9
- LED lighting... 9
- lithium ion battery... 40

M
- magnetic south... 50
- maximum power point tracking... 57
- modified sine wave inverters... 25
- monocrystaline.. 44-45

N
- National Electrical Code... 15, 62
- National Renewable Energy Lab... 51
- nickel cadmium battery.. 40
- nominal voltage... 33, 39, 71, 75

O

- overcurrent protection.......................... 61, 64, 65, 68, 70-71, 78

P
- parallel.. 38-39
- peak sun hours.. 50
- polarity.. 27, 79, 92
- polycrystaline... 44-45
- power equation.. 30-31, 36
- pulse width modulation... 56, 57
- PV module... 12
- PVWatts... 51

S
- sealed batteries.. 40-41
- series... 38-39, 46, 56-57
- solar panel 11-12, 41, 43-52, 57-59, 66-69, 84-89, 93, 96, 102, 108-109, 111, 114
- solar south... 49-50
- stand-alone photovoltaic system... 11, 29
- standard test conditions (STC).............................. 46, 58, 67, 86
- surge capacity.. 22, 23, 70-73, 75

T
- thin film.. 44-45
- true south.. 50

V
- volts 4, 12, 15, 18-20, 30-31, 33-34, 36, 39, 51-53, 56-57, 62, 70-73, 75, 83, 85, 90-91, 105, 112, 114

W
- watt-hour... 4, 30

watts 4, 12, 18-24, 30, 35-36, 44-46, 50-51, 67, 70-73, 81-83, 85-86, 89-91, 98, 102, 105
waveform.. 24-26

www.ingramcontent.com/pod-product-compliance
Lightning Source LLC
Chambersburg PA
CBHW050559300426
44112CB00013B/1988